MEDICAL
REVIEW
SERIES

Biochemistry

Biochemistry

Second Edition

Compiled and Written by
Nikos M. Linardakis, M.D.
and
Christopher R. Wilson, M.D.

McGRAW-HILL
Health Professions Division

New York St. Louis San Francisco Auckland Bogotá Caracas Lisbon
London Madrid Mexico City Milan Montreal New Delhi San Juan
Singapore Sydney Tokyo Toronto

McGraw-Hill

A Division of The McGraw·Hill Companies

BIOCHEMISTRY: Digging Up the Bones

1234567890 MAL MAL 9987

ISBN 0-07-038217-4

This book was set in Times Roman by V & M Graphics, Inc.
The editors were John Dolan and Steven Melvin;
the production supervisor was Helene G. Landers;
the cover designer was Matthew Dvorozniak.

Malloy Lithographing, Inc., was the printer and binder.

Cataloging-in-Publication Data is on file for this title
at the Library of Congress

To my father, Robert E. Wilson, M.D.—C.R.W.

and

To Roger S. Peterson and Mrs. Lee, in memoriam—N.M.L.

Contents

Preface

Prepare to learn! The *Digging Up the Bones Medical Review Series* is a compilation of concepts frequently seen in course exams and on USMLE Step I. This volume offers a clear, concise, and to-the-point presentation of the facts you need to succeed in Biochemistry. The *Digging Up the Bones* review series is designed to complement your course work, providing in a brief format items you need to review in a subject. Pay particular attention to **bold** or *italicized* words. We recommend placing notes in the margins and reviewing the material at least two times. Try to make use of any old exams that are available to you, these are sometimes carried in your library, by other students in the year ahead of you, or through the professors who teach the course. A final tip: Use this book to preview material that will be covered on the next day of class and you will be surprised with how much more you get out of lectures.

Best of luck.

DIGGING
—UP THE—
BONES®

MEDICAL
REVIEW
SERIES

Biochemistry

General Biochemistry

ENERGY

Energy from the sun is stored in "high energy" bonds when CO_2 and H_2O are combined by photosynthesis to form carbohydrates. Ultimately it is this energy which drives all the reactions necessary to maintain life on this planet.

Photons (from sunlight) ⟶ excited electrons in chlorophyll

$CO_2 + H_2O$

photosynthesis

high energy carbon-carbon bonds (carbohydrates)

catabolic metabolism (enzyme linked oxidation-reduction)

⟶ CO_2

electrons (2H in NADH or $FADH_2$) ⟵ in higher organisms

O_2

⟶ H_2O

electron transport chain

high energy phosphate bonds (ATP)

anabolic metabolism + enzymes + substrates (building blocks)

complex biologic molecules

MATTER

Plants and bacteria also extract nitrogen and CO_2 from the atmosphere, minerals from the soil, and synthesize molecules which become the building blocks of biomatter. Other organisms consume and incorporate these building blocks into their substance by the processes of growth, maintenance, and reproduction.

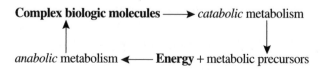

Complex biologic molecules ⟶ *catabolic* metabolism

anabolic metabolism ⟵ **Energy** + metabolic precursors

As these processes are repeated, energy and matter are continually consumed and recycled throughout the chain of life. Life produces and reproduces itself by the processes of replication, transcription, and translation, using the instructions of the genetic code stored in DNA.

1. Replication:
   $$DNA \longrightarrow DNA$$

2. Transcription:
   $$DNA \longrightarrow RNA$$

3. Translation:
   $$RNA \longrightarrow Proteins$$

ACID-BASE

H+:   Hydrogen ion.
[ ] :   Concentration, i.e., $[H^+]$ = concentration of hydrogen ions.
pH: $-\log [H^+]$.
Base: H^+ acceptor.
Acid: H^+ donor.
Weak Acid: donates few of its H^+.
Strong Acid: donates almost all of its H^+ and forms a weak conjugate base.
K: used to symbolize a constant.
K_a: constant which correlates with the extent an acid gives up its H^+.
   (the stronger the acid the higher the K_a)
pK_a: $-\log K_a$

HENDERSON-HASSELBACH EQUATION

H^+:   hydrogen ion.
A^-:   conjugate base.
HA: An undissociated acid.

$$K_a = \frac{[A^-] \times [H^+]}{[HA]} = \frac{\text{conjugate base concentration} \times \text{hydrogen ion conc.}}{\text{undissociated acid conc.}}$$

If you solve this equation for [H+]:

$$[H+] = K_a \times \frac{[HA]}{[A-]}$$

Now if you remember how to use Logarithms, you can rearrange this equation:

$$-\log [H+] = -\log K_a + (-\log [HA] / [A-])$$

but remember $-\log A/B = +\log B/A$.

So you get: $$pH = pK_a + \log \frac{[A]}{[HA]}$$ *Which is the **Henderson-Hasselbach** equation!*

BUFFER

A solution that resists changes in its pH when acid or base are added. Contains acid-conjugate base pairs, buffering is greatest when pH = pK, and when this happens, the concentration of acid and its conjugate base are equal (i.e., $[A^-] = [HA]$). The proton release or acceptance is directly proportional to the pH.

MAXIMUM BUFFERING CAPACITY

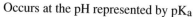

Occurs at the pH represented by pK_a

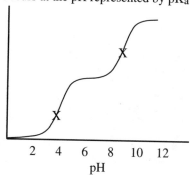

pH

TYPES OF TRANSPORT

Osmosis is the migration of water from the side of a membrane that is hypo-osmotic to the side that is hyper-osmotic.

Facilitated transport uses carriers:

Passive transport is the transport of a molecule "down the concentration gradient" by a carrier; does not use energy.

Active transport is the transport of a molecule (i.e., Na^+, Glucose, Amino acid) by a carrier which uses energy.

Co-transport, a type of active transport, transports an ion along with a molecule as in the case of the glucose-Na^+/K^+ ATPase pump. This system uses energy from ATP to *co*-transport sodium and glucose from intestinal epithelial cells into the blood.

Metabolism   2

METABOLISM

The process in which matter from outside an organism is transformed into energy or material for the organism. In bacteria, metabolism takes place inside the cytoplasm and along the cell membrane. In higher organisms, metabolism occurs in the cytoplasm and in the mitochondria.

CATABOLISM
Reactions that break the bonds of complex molecules. In the body, these reactions may be linked to enzyme catalysts allowing for storage of the released energy in the form of ATP (adenosine triphosphate). Polymers yield energy when they are converted to intermediates.

ANABOLISM
Reactions which synthesize complex molecules from simpler molecules using ATP as an energy source. Simple molecules become polymers.

ATP (Adenosine Triphosphate)
Energy medium for reactions inside the cell.

Oxidation   Loss of electrons
Reduction   Gain of electrons

CALORIE (C)   In nutrition, equals one scientific kilocalorie (Kcal).

BOMB CALORIMETER
Insulated chamber in which samples of food are oxidized and the amount of heat generated is accurately measured.

Energy Source	Energy Utilized in Humans
Carbohydrates	4 Kcal/gram
Proteins	4 Kcal/gram*
Alcohols	7 Kcal/gram
Fats	9 Kcal/gram

*Protein would yield 5 Kcal/gram if it were completely metabolized but this would produce compounds too toxic to the body. Instead urea is the ultimate excretory form and still has some energy left in it which can be utilized by bacteria. Bacteria can metabolize urea to ammonia. Ethanol is metabolized in the liver. Alcohol inhibits gluconeogenesis.

4

$$\text{Ethanol} \xrightarrow[\text{Alcohol dh}]{\quad} \text{Acetaldehyde} \xrightarrow[\text{Acetaldehyde dh}]{\quad} \text{Acetate}$$

with $NAD^+ \to NADH$ and $NAD^+ \to NADH$ across each reaction.

NAD^+ = The oxidized form; the electron acceptor
NADH = The chemically reduced form; the electron donor.
Each reaction above is a reduction reaction = the gain of an electron (H^+).

ENERGY STORAGE

Energy sources are oxidized to produce energy. Excess Calories are stored in the body. In nutrition, the "70 Kg Man" is the defined standard with the following composition:

Amount	~% Of Stored Energy
15 Kg Adipose (Fat)	85%
6 Kg Protein	14.5%
.25 Kg Carbohydrates	0.5%*

*glycogen stored in the muscle and liver

GLYCOGEN

A starch made in the liver and muscle. It is the first fuel used to power activity. Carbohydrates are NOT a cause of any diseases—the only problem may be adding dental caries.

FAT

Fat (triacylglycerol) is the major energy store in the body. Adipose tissue is an efficient way to store energy because fat is twice as energy dense as carbohydrate or protein, and has less water associated with it. When excess calories are taken in, only limited amounts of carbohydrates and protein can be stored (unless they are converted to fat which uses energy). Fats need little processing and are easily stored; this is one reason too much fat in the diet can be a problem. The current percentage of fat in the U.S. diet is near 42 percent, and the goal is to reduce the fat calories to ≤ 30 percent of the total calories (fat, carbohydrates and protein intake). Olive oil and Canola oil are mono **un**saturated fats. All plants have **un**saturated fat (contain double bonds and are more fluid), except coconut and palm oil, which are *saturated* fats and increase the risk for cardiovascular disease.

PROTEIN

Protein is not metabolized as an energy source in healthy, well-nourished people, but there is some turnover due to repair and renewal of tissue. The biologic value of *animal* proteins is highest compared to plants (which may be missing several essential amino acids). This is why the lowest fat and inexpensive form of protein sources would be a combination of animal and plant derived foods.

Tissue	Energy Sources Utilized
Muscle	at rest: fat
	when active: carbohydrates
Brain	carbohydrates

RESPIRATORY QUOTIENT (R.Q.)
CO_2 made and O_2 used by a tissue when a given type of energy source is oxidized.

$$R.Q. = \frac{CO_2 \text{ produced}}{O_2 \text{ consumed}}$$

R.Q.'s when one energy source is oxidized	
Fat	0.7
Protein	0.8
Carbohydrate	1.0

A body at rest will have an overall R.Q. of ~0.8 because resting muscle is utilizing fat, brain is using carbohydrates, and there is some turnover of protein. As a body becomes active the R.Q. approaches 1.0 since muscle begins to use carbohydrates.

ENERGY REQUIREMENTS
BMR + Physical Activity + SDA = Total Energy Requirement (TER)

SDA (specific dynamic action): The amount of energy used secondary to the increase in the metabolic rate that occurs during the digestion and absorption of food. This small number is often left out of calculations. (Thermic effect of food is about 10% of the energy consumption.)

Physical Activity: Energy used by physical activity. In a sedentary person accounts for about 15–25% of the TER, in a very active person can be as much as 50%. (Physical activity is about 30% of the energy consumption.)

BMR (Basal Metabolic Rate) = TER − [SDA + Physical Activity] BMR is the energy required for the maintenance functions of the body, i.e., energy needed for the brain, respiration, and to maintain homeostasis. (This is about 60% of the energy expenditure of a resting person.)

BMR can be estimated as 24 Kcal/kg for adult males and 18 Kcal/kg for adult females (per 24 hours), or 24 kcal/kg × 70 kg male = 1,680 kcal/day minimum caloric necessity.

Amino Acids and Proteins

3

AMINO ACIDS

There are 20 amino acids in animal metabolism, but more occur in nature. Amino acids, the building blocks of proteins, have different side chains which are connected to a common backbone. At physiologic pH, all amino acids have both positive and negative charges.

Structure of amino acid:

$$+H_3N —\overset{\overset{\textstyle H}{|}}{\underset{\underset{\textstyle R}{|}}{C}}—COOH$$

$$R = \text{side chain}$$
$$NH_3 = \text{amino group}$$
$$COOH = \text{carboxyl group}$$
$$-\overset{|}{\underset{|}{C}}- = \text{alpha carbon (}\alpha\text{-carbon)}$$

When the pK is small, then the proton comes off easily (readily). When the pK is very large, then it is difficult to remove the proton, and you must add a base (titrate) and remove the protons.

Side Chains:

Nonpolar   **Structure**   **Simplified Abbreviation**
(Electrons are *evenly* distributed)

Glycine Gly G	–H	–H
Alanine Ala A	$-CH_3$	–C

Side Chains:

Nonpolar	Structure	Simplified Abbreviation

(Electrons are *evenly* distributed)

Valine
Val
V

$-CH$
CH_3 CH_3

$-C$
C C

Leucine
Leu
L

$-CH_2-CH$
CH_3 CH_3

$-C-C$
C C

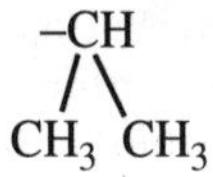

Isoleucine
Ile
I

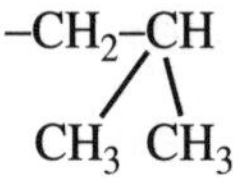

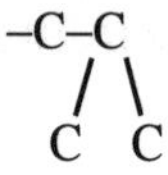

$-CH$
CH_3 CH_2
CH_3

$-C$
C C
C

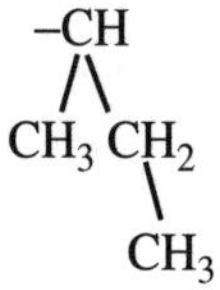

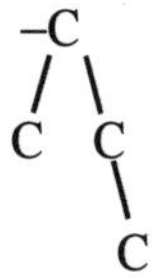

Methionine
Met
M

$-CH_2-CH_2-S-CH_2$

$-C-C-S-C$

Phenylanine
Phe
F

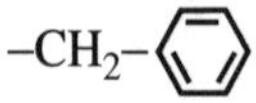

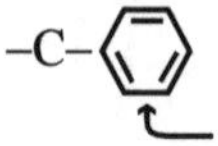

$-CH_2-$⬡

$-C-$⬡
└── aromatic side chain

Tryptophan
Trp
W

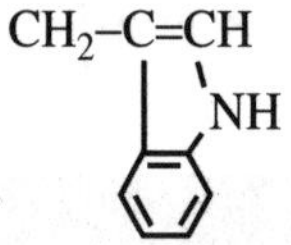

$CH_2-C=CH$
NH

C
N

Proline
Pro
P

$^+H_2N-\overset{\displaystyle H}{\underset{\displaystyle}{C}}-COO^-$
H_2C CH_2
CH_2

$^+H_2N-C-COO^-$
C

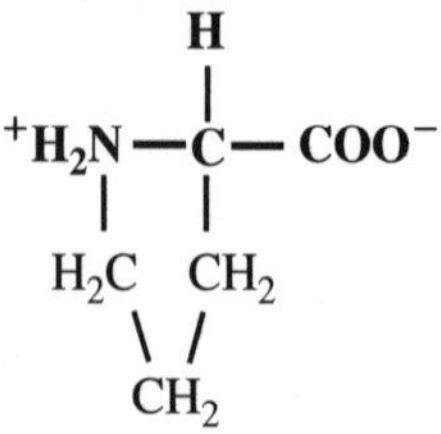

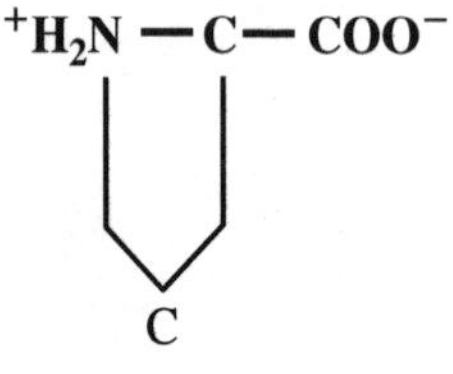

Proline is an imino acid with a side chain that forms a ring with the amino group of the **backbone**.

Valine, Leucine, and Isoleucine are hydro**phobic** and bond together through hydrophobic interaction. (Imagine oil and water: Val, Leu, and Ile are the oil and they try to avoid water.)

Physiologic pH = 7.4

Side Chains:

Polar (Electrons are **un**evenly distributed)	Structure	Simplified Abbreviation
Serine Ser S	$-CH_2-OH$	$-C-OH$
Cysteine Cys C	$-CH_2-SH$	$-C-SH$
Threonine Thr T	$-CH-CH_3$ | OH	$-C-C$ | OH
Tyrosine Tyr Y	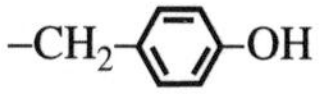 $-CH_2-$⟨◯⟩$-OH$	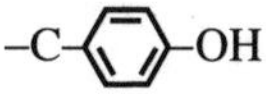 $-C-$⟨◯⟩$-OH$

Polar Amide

Asparagine Asn N	$-CH_2-\overset{\|}{\underset{\|}{C}}-NH_2$ O	$-C-\overset{\|}{\underset{\|}{C}}-NH_2$ O
Glutamine Gln Q	$-CH_2-CH_2-C-NH_2$ O	$-C-C-C-NH_2$ O

Polar Acidic

Aspartic Acid Asp D	$-CH_2-C-OH$ O	$-C-C-OH$ O
Glutamic Acid Glu E	$-CH_2-CH_2-C-OH$ O	$-C-C-C-OH$ O

Side Chains:

Polar	Structure	Simplified Abbreviation

(Electrons are **un**evenly distributed)

Continued

Polar Basic (+1 Charge)

Lysine $-CH_2-CH_2-CH_2-CH_2-NH_3+$ $-C-C-C-C-NH_3+$
Lys
K

Arginine $-CH_2-CH_2-CH_2-NH$ $-C-C-C-NH$
Arg
R

$$\begin{array}{cc} \quad\quad\quad | \quad\quad\quad\quad\quad & | \\ \quad\quad\quad C \quad\quad\quad\quad\quad & C \\ \quad\quad / \;\backslash\!\backslash \quad\quad\quad & / \;\backslash\!\backslash \\ H_2N \quad NH_2+ \quad\quad & H_2N \quad NH_2+ \end{array}$$

Histidine $-CH_2-C-NH^+$ $-CH_2-C-NH^+$
His
H

$$\begin{array}{cc} \quad\quad || \quad || \quad\quad & || \quad || \\ \quad\quad HC \quad C \quad\quad & C \quad C \\ \quad\quad \backslash \; / \quad\quad\quad & \backslash \; / \\ \quad\quad N \quad\quad\quad\quad & N \\ \quad\quad | \quad\quad\quad\quad & | \\ \quad\quad H \quad\quad\quad\quad & H \end{array}$$

Note: • Phenylalanine, Tyrosine and Tryptophan have **aromatic** side chains.
      • Cysteine and Methionine contain **sulfur**.
      • Proline is an **imino** acid and forms a ring with its own backbone.
      • The other 19 are **amino** acids.

SIGNIFICANCE OF SIDE CHAIN CHARACTERISTICS

Side chains are what ultimately determine the function of the protein that they make up.

Non-Polar Side Chains: Are hydrophobic (fear water), so regions of a protein with many non-polar amino acids will try to stay away from water. This tendency may contribute to the three dimensional shape of a protein.

Tyrosine has a polar side chain but it is still somewhat hydrophobic because of its aromaticity.

Charged side chains: Aspartate (−), Glutamate (−), Lysine (+), Arginine (+), Histidine (+) may take part in ionic interactions.

Histidine has an imidazole ring; it also offers buffering at physiologic pH to protein. Histidine is decarboxylated to histamine (a vasodilator).

Serine and threonine contain hydroxyl groups which can form hydrogen bonds.

Cysteine:
- The sulfhydryl group is an important part of the active site of many enzymes.
- In addition, two side chains may form a covalent disulfide bond which may join two different proteins or two different regions of the same protein.

$$^+_3HN-\overset{\overset{\displaystyle H}{|}}{C}-CH_2-S-S-CH_2-\overset{\overset{\displaystyle H}{|}}{C}-NH_3{}^+$$
$$\underset{\displaystyle COO^-}{|}\qquad\qquad\underset{\displaystyle COO^-}{|}$$

A disulfide bond (Cystine)

Cystine is formed by a disulfide bond joining of two cysteine residues. Keratin contains alot of cystine.

Glycine is used in first step of heme synthesis:

Glycine + succinyl CoA ⟶ δ-ALA (δ-aminolevulinic acid).

Lysine has an aliphatic side chain and a highly basic amino group at physiologic pH. Blood glycoproteins in diabetics contain glucose linked to lysine.

Tryptophan has the largest side chain; it is hydrophobic with an aromatic ring. Tryptophan deficiency can cause Hartnup disease and Pellagra. (Trp ⟶ Niacin) Tryptophan is a precursor of Serotonin (5-HT). Tryptophan is *hydroxylated* to 5-Hydroxytryptophan, and *decarboxylated* to **5-H**ydroxytryptamine.

Alanine carries nitrogen from peripheral tissues to the liver.

Proline disrupts an α-helix in a polypeptide. Proline is usually the residue at the β-turn in β-pleated sheets (see secondary structure).

Serine is the phosphorylation site of enzyme modification. Serine may be dehydrated and deaminated *directly*. In glycoproteins, serine is often linked to the carbohydrate group.

Glutamine is deaminated by glutaminase resulting in the formation of ammonia. It is a major carrier of nitrogen to the liver (from peripheral tissues).

ZWITTERIONS
Dipolar ions. Amino acids have a dipolar (+ and −) charge at neutral pH. From the protonated amino group (NH_3^+), and ionized carboxyl group (COO^-).

OPTICAL ACTIVITY
Polarized light passing through an optically active sample is rotated to the left or right. The alpha-carbon of all amino acids, except glycine, is bound to 4 different groups so it is called "**chiral**" and this makes amino acids *optically active.*

CONFIGURATION
Amino acids are in 2 configurations: "**L**" or "**D**." In man, they are usually **L**-amino acids (remember, **L**-amino acids and **D**-sugars).
[L = levorotatory ; rotation to the left]
[D = dextrorotatory ; rotation to the right]

(dexter = Latin word for right)

Higher organisms contain only **L**-amino acids in their proteins, but **D**-amino acids do occur in the peptidoglycans of bacterial cell walls.

ESSENTIAL AMINO ACIDS
1. Phenylalanine
2. Valine
3. Tryptophan
4. Threonine
5. Isoleucine
6. Methionine
7. Leucine
8. Lysine

RELATIVELY ESSENTIAL AMINO ACIDS*
1. Histidine
2. Arginine

These amino acids are not essential in the short term. A small amount of Arginine is made in the body, but not enough for growing children.

Histidine is not made at all.
It is recycled, but must eventually be consumed.

ESSENTIAL AMINO ACIDS (may be easier for you to memorize these only)
These are the amino acids whose intake is essential for proper body function—since they can**not** be synthesized.
Remember "Private Tim Hall" (PVT TIM HALL):
Phenylalanine, **V**aline, **T**ryptophan,
Threonine, **I**soleucine, **M**ethionine,
Histidine*, **A**rginine*, **L**eucine, **L**ysine.

NON-ESSENTIAL AMINO ACIDS

Amino acids that can be synthesized in sufficient amounts in humans or that may be made from precursors.
i.e., Arginine (in adults), tyrosine (hydroxyl group on Phe), cysteine, proline.

BIOLOGIC VALUE

The relative ability of a protein source to provide all the essential amino acids. Animal sources have the highest biologic value while individual vegetable sources are lower (because they lack enough of one or more of the essential amino acids). However, combined different vegetable sources can result in meals of high biologic value that provide enough of all the essential amino acids.

NITROGEN BALANCE

Nitrogen balance is a measure of the protein status of the body. Most of the nitrogen in the body is from consumption of amino acids. When not enough protein is ingested, the body breaks down its own protein and excretes more nitrogen than what is taken in; the body is in *negative nitrogen balance.* On the other hand, growing children with adequate protein intake will store more nitrogen in their new muscles than they excrete and will be in *positive nitrogen balance.*

BUFFER REGION

This is the area where the least amount of change in the pH occurs with the addition of a base.

Titration Curve

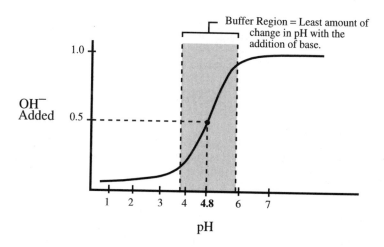

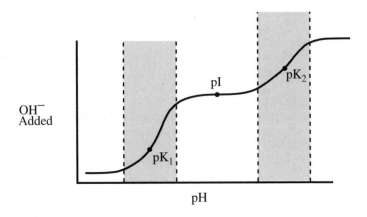

pH = pKa at midpoint

$$HA \xrightleftharpoons[H^+]{OH^- \quad H_2O} A^-$$

(–COOH)   (COO⁻)

Diseases of Malnutrition in Children:

Kwashiorkor
This disease is due to an inadequate intake of protein, but an adequate intake of calories. Symptoms: Unpigmented hair which appears reddish, fatty liver, edema due to low serum albumin producing a protruding abdomen ("pot belly").

Marasmus
Disease due to an inadequate intake of calories *and* protein. Total calorie deficiency and protein deficiency. Symptoms: Retarded growth, emaciation, "cachexia", but **no** low albumin or edema.

PROTEINS

The Peptide Bond
Amino acids are joined linearly by peptide bonds which are formed by a reaction called **dehydration**. One molecule of H_2O is produced per peptide bond formed. The peptide bond has a **partial double bond character** and is rigid.

Peptides
2 amino acids joined by a peptide bond are called a **di**peptide; many amino acids joined by peptide bonds are called **poly**peptides. By convention polypeptides are written from the N-terminal to the C-terminal. (Amino end to carboxyl end—remember A before C)

Peptide Bond

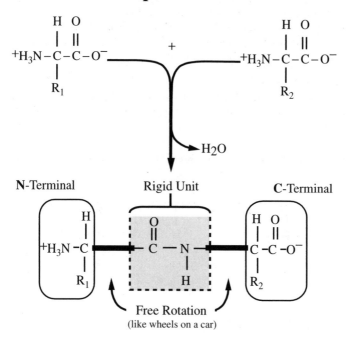

N-Terminal

Rigid Unit

C-Terminal

Free Rotation
(like wheels on a car)

STRUCTURE OF PROTEINS

A protein may consist of a single polypeptide chain or more than one chain. The ultimate shape of a protein is the result of the following:

1. PRIMARY STRUCTURE

Sequence of amino acids in polypeptide chain, i.e.:

N-leu-met-glu-val-ala-leu-gly-gly-phe-C

The primary structure is important for deciding the higher structure of proteins. It is simply the *amino acid sequence*. It also determines the location of disulfide bonds.

2. SECONDARY STRUCTURE

If present, it is the result of hydrogen bonding between the C=O of one peptide bond and the N–H of another peptide bond. This hydrogen bonding produces regularly repeated structures, the most important of which are the α-helix and β-pleated sheets. However, these structures are **not** present in all polypeptides.

α-Helix

The α-helix is stabilized by hydrogen bonds. Hydrogen bonds are between a carboxyl and an amino group, and occur four amino acids away in the same polypeptide chain (actually 3.6 amino acids per turn). Hydrogen bonds are *parallel* to the axis of the helix, and the R-groups come off the sides. The α-helix is a rod-like structure and in man is usually a *right*-handed helix. The side chains point away from the center of the rod. An a-helix is disrupted by *Proline*. Regions with many charged or bulky side chains (i.e., tryptophan) can also disrupt the helix.

β–Pleated Sheets
There are two types of β-Pleated Sheets, *Parallel sheets* and *Anti-parallel sheets*. Hydrogen bonds occur between 2 different polypeptide chains or 2 regions of the same chain.

Parallel sheets run in the same direction as compared to their terminal ends:

Anti-parallel sheets run in opposite directions:

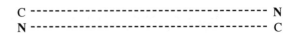

3. TERTIARY STRUCTURE
The overall 3-dimensional shape of a protein. Most proteins are *globular*. Some proteins are *fibrous*, like collagen. The tertiary structure is the result of the combination of the primary structure, secondary structure and the interactions between the different side chains, i.e.:

* Covalent disulfide bonds between 2 cysteines. This anchors two chains and is found in proteins designated for export.
* Hydrophobic side chains are oriented to the *inside* of globular proteins and *away* from water.
* Hydrogen bonds between the side chains of threonine and serine.
* Ionic interactions between the charged side chains.

Remember:
Hydro*philic* amino acid side chains are usually located on the **out**side of proteins. Hydro*phobic* amino acid side chains are usually on the **in**side. Charged amino acid side chains are usually located on the **out**side of proteins. Covalent bonds in proteins also include the peptide bond and the sulfhydril bond.

Domains:
Tertiary structure may result in separate highly organized units within the same polypeptide often with distinct functions, i.e., an active site or binding site of an enzyme.

Bends may form from the steric relationship of amino acids away from each other.

4. QUATERNARY STRUCTURE
Some proteins are made up of multiple subunits, each of which contain a single polypeptide chain. These subunits join to form a single protein held together by the same forces that maintain tertiary structure. The number of subunits and their spatial arrangement comprise quaternary structure. In general, *several* polypeptides make a protein *functional*. Examples include: Hemoglobin and Myoglobin.

DETERMINATION OF AMINO ACID SEQUENCE

HYDROLYSIS
The dehydration reaction that forms the peptide bond can be reversed by a reaction called *hydrolysis*. In a strong acid solution at 43°C (110°F) for 24 hours all the peptide bonds of a polypeptide chain will be broken and the individual amino acids released.

A sample containing one type of polypeptide is hydrolyzed then placed in an amino acid analyzer, which uses 2 steps to identify and quantitate the amino acids in the sample:

1. Ionic Exchange Chromatography
In an acidic solution, amino acids have a *positive* charge and become bound to a negatively charged resin in the chromatographic apparatus. Then, solutions of varying pH and ionic strength cause the amino acids to become negatively charged and released from the resin. Each type of amino acid is identified by the specific combination of pH and ionic strength at which it is released. Then each of the different types of amino acids goes onto the next step. This tells you which amino acids are in the original polypeptide.

2. Spectrophotometry
Amino acids are labeled with *ninhydrin* using heat. A spectrophotometer measures the absorbance of each labeled amino acid. The amount of light absorbed by each type of amino acid is proportional to the amount of that amino acid in the original polypeptide (i.e., the most abundant amino acid absorbs the most light, amino acids present in the same proportion absorb the same amount of light).

Absorption of light at 280 nm: Trp > Tyr > Phe > Leu

It is possible to determine the amino acid sequence (primary structure) of a polypeptide by using the following methods:

N-terminal amino acid:

1. Sanger's method
Fluorodinitrobenzene labels the free amino group on the N-terminal of a polypeptide with a dinitrophenyl (DNP) group. Then, acid is used to hydrolyze away the remaining amino acids in the chain and the labeled N-terminal amino acid is identified using ion exchange chromatography.

2. Edman's Reagent
Also labels the free amino group on the N-terminal, but the labeled amino acid can be removed from the rest of the polypeptide chain and the process repeated until the entire amino acid sequence of a small protein can be determined.

Therefore, Edman degradation removes one amino acid at a time, sequentially. Phenyl isothiocyanate is used in this process.

C-terminal amino acid:

1. Hydrazine
Binds to the −COO⁻s in the peptide bonds but not to the −COO⁻ at the end of the C-terminal; this last amino acid can be released and identified.

2. Carboxypeptidase
An enzyme that cleaves the peptide bonds sequentially starting at the C-terminal. Carboxypeptidase is a *zinc* protease.

Sequencing of Longer Polypeptides
Larger polypeptides can be cleaved into smaller fragments, and the sequence of these fragments can be determined. By using different agents (to generate fragments that overlap), the entire sequence of larger proteins can be determined.

Cyanogen Bromide
Cleaves polypeptides at the carboxyl side of *methionine.*

Trypsin
Digestive enzyme produced by the pancreas which cleaves polypeptides on the carboxyl side of either *arginine or lysine.* (Acts on the basic amino acids with cationic amino side chains.)
Acts as a serine protease.

Chymotrypsin
Unreliably cleaves polypeptides at the carboxyl side of *tryptophan, tyrosine,* or *phenylalanine* (aromatic amino acids). Acts as a serine protease (most pancreatic enzymes are serine proteases).

Proteins Composed of Multiple Polypeptides
The quaternary structure of some proteins are composed of subunits made up of more than one kind of polypeptide chain. **Denaturing agents** disrupt the

forces that maintain tertiary and quaternary structure. This separates the different polypeptides so they can be sequenced. The 2 common denaturing agents are: **Urea** and **Guanadine hydrochloride**. After the individual polypeptide chains are separated they may undergo amino acid analysis.

ENZYMES

ENERGY OF ACTIVATION
The amount of energy which must be added to a reaction to allow it to go forward. Enzymes (catalysts) *decrease* the energy of activation.

Energy of Activation

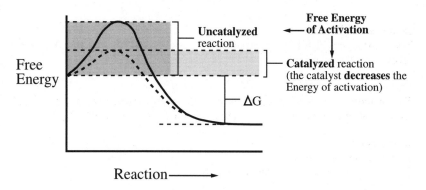

REACTION VELOCITY
This is the *rate* that a reaction occurs.

CATALYSTS
Catalysts increase the *speed* of a reaction by *lowering* the reaction's energy of activation.

ENZYME
A protein catalyst, which may require a vitamin or mineral cofactor. Enzymes may increase the rate of a reaction between a thousand and a million times the uncatalyzed rate. Enzymes also lower the temperature at which reactions take place to a temperature compatible with life. Enzymes catalyze reactions by decreasing the free energy of activation (Gibb's free energy of activation). The active site is a small crevice which initially weakly binds substrates. Metal ions may act as cofactors by keeping an enzyme in its active conformation.

Isomerase: converts one isomer to another
Epimerase: converts one epimer to another
Carboxylase: adds carboxyl group to a molecule

Kinase: catalyzes reactions adding or removing nucleoside triphosphates (*usually* ATP).
Dehydrogenase: removes H^+

ENZYME-SUBSTRATE COMPLEX
The intermediate product of an enzyme catalyzed reaction.

k_1, k_2, k_3 = Rate Constants

$$E + S \underset{k_2}{\overset{k_1}{\rightleftharpoons}} ES \xrightarrow{k_3} E + P$$

$$Km = \frac{k2+k3}{k1}$$

SUBSTRATE
The substance which is recognized by the enzyme and is transformed into the product by the reaction.

PRODUCT
The product is the substance formed by the interaction of the substrate and enzyme. Removal of the product helps the reaction to proceed.

ACTIVE SITE
1. binds the specific substrate
2. the part of the enzyme which is catalytic.

AFFINITY
Attraction between enzyme and substrate

SPECIFICITY
The binding site of an enzyme may be so "specific" that it recognizes and binds to only one optical isomer of a given molecule; or may be less specific and recognize several closely related molecules.

Lock and Key Theory
Substrate and enzyme *always* fit each other.

Induced Fit Theory
Substrate and enzyme *fit only at binding.*

ISOENZYMES
Enzymes which have different amino acid sequences, but catalyze the same reactions. (Isoenzymes have different electrophoretic mobilities and are made of different proteins.) For example, **LDH** or **L**actate **d**ehydrogenase or **H4**—which is found in the **h**eart and kidney, and **M4**—which is found in liver and skeletal **m**uscle.

HOLOENZYME
An enzyme which requires a co-factor to be active.

APOENZYME
Protein part of a holoenzyme.

CO-FACTOR
Vitamin or mineral. Apoenzyme + Cofactor = Holoenzyme

PROSTHETIC GROUP
A co-factor which is permanently complexed with its enzyme.

ALLOSTERIC ENZYME
An allosteric enzyme contains another site, different from the active site, at which an effector binds. *Positive* effectors *increase* the rate at which the enzyme will catalyze a reaction, *negative* effectors *slow* the rate. These enzymes are often the first step or branch point in a metabolic pathway, with the product of the pathway serving as a negative effector and excess substrates as positive effectors.

Homotropic enzyme substrate molecules: the binding of substrate causes enzymes to have a conformational change, which increases the binding of other substrate molecules and the velocity. The *homotropic* enzyme's *substrate* is also its effector.

Heterotropic enzyme substrate molecules: different ligand interactions. A *positive* heterotropic effector is an activator. A *negative* heterotropic effector is an inhibitor. (This does NOT change the equilibrium of the reaction.) The *Heterotropic* enzyme has a different molecule as an effector (i.e., the end *product* of the pathway).

ALLOSTERIC ENZYME
An allosteric enzyme may increase or inhibit substrate binding. Substrate binding to one site can alter other sites. The plot of substrate concentration vs. velocity is *sigmoidal,* and does NOT follow Michaelis-Menten kinetics (which is a hyperbolic plot). (*See plots*)

PHOSPHORYLATION
Activating or inactivating regulatory enzymes. The reaction is catalyzed (or activated) by phosphorylase kinase. For example, inactive *phosphorylase b* is activated to phosphorylase *a*. It is deactivated by *phosphatase.*

FACTORS THAT AFFECT THE RATE OF A REACTION:

1. SUBSTRATE CONCENTRATION [S]
Increasing the substrate will increase the reaction rate until all of the enzyme binding sites are occupied, at the Maximal Velocity (Vm). Remember, the enzyme is like a door—only so many people can go through at once.

2. TEMPERATURE

Increasing the temperature increases the rate because more molecules can achieve the energy of activation, until the increased temperature causes the protein in the enzyme to either *deform* to a conformation that is not catalytic, or to *denature*.

3. pH

Different enzymes have different optimal pH values for a maximal reaction rate.

KINETIC ORDER

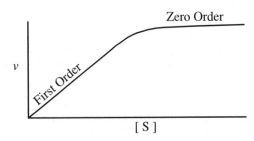

Zero Order

Reactions where the rate is independent from substrate concentration.

$$V = K[S]^0$$

First Order

Reactions where the rate is dependent on the substrate concentration; the *rate decreases* over the time as the substrate concentration decreases. The rate is proportional to the substrate concentration.

$$V = K[S]^1$$

Second Order

$$V = K[S]^2$$

V = velocity or rate of reaction, [S] = substrate concentration, K = rate constant.

MICHAELIS-MENTEN EQUATION

Enzymes that follow Michaelis-Menten kinetics have a *hyperbolic* curve (unlike allosteric enzyme reactions, which have a *sigmoid* curve). Assumes:
1. [S] much greater than [E], so all binding sites are filled
2. [ES] is constant
3. [P] is low

$$v = \frac{Vm[S]}{Km + [S]}$$

v   = velocity
Vm = maximal velocity at saturation
Km = Michaelis Constant; the substrate concentration which gives half the Vm.
k_1, k_2, k_3 = Rate Constants.

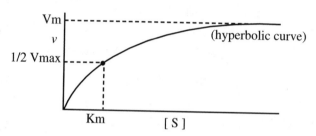

$$Km = \frac{k2+k3}{k1}$$

Michaelis-Menten Plot

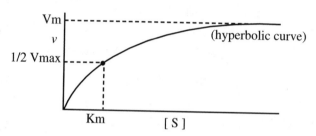

When v = 1/2 Vm, then Km= [S]
Higher enzyme-substrate *affinity* = *Lower* Km
A *lower* affinity of the enzyme for the substrate means a *larger* Km.

LINEWEAVER-BURKE PLOT

The Lineweaver-Burke equation rearranges the Michaelis-Menten equation so that the line plotted is straight which makes finding Vm easier.

$$\frac{1}{v} = \frac{Km}{Vm[S]} + \frac{1}{Vm}$$

Lineweaver-Burke Plot

for example

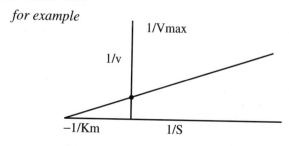

INHIBITION

A. REVERSIBLE

1. Competitive

Mechanism: Inhibitor is shaped similar to the substrate and temporarily competes with it for binding sites.

Reversed: By increasing [S] (substrate concentration)

Km: Higher [I] (inhibitor concentration) increases Km-simulating decreasing enzyme-substrate affinity.

Vm: Not changed

Competitive inhibitor of enzyme that *increases* K_m, but has NO effect on V_{max}.

Competitive Inhibition

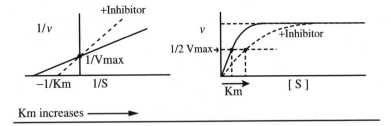

Km increases ⟶

2. Non-Competitive

Mechanism: Binds temporarily to enzyme somewhere other than active site but halts catalysis. Has the same effect as reducing [E] (enzyme concentration)

Reversed: By increasing [E]

Km: Not changed

Vmax: Lowered

Non-competitive inhibition decreases V_{max}, and has NO effect on Km. This inhibitor binds to both enzyme and Enzyme-Substrate complex.

Non-competitive Inhibition

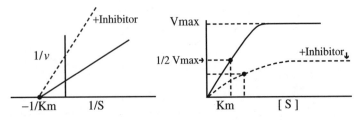

B. IRREVERSIBLE

Irreversible inhibitors (Non-competitive inhibitors)
Bind covalently to enzymes and permanently inactivate them (i.e., organo-phosphates, used in nerve gas and insecticides, bind permanently to acetylcholinesterase) are not plotted on Michaelis-Menten graphs.

Allosteric Enzymes
Usually these are multimeric (Hb, $\alpha 2\beta 2$). They catalyze irreversible reactions and are usually the committed step in the pathway—and the slowest step of the reaction (like HMG CoA reductase). These enzymes are found at metabolic brach points.

Non-Michaelis-Menten
(Allosteric Enzyme, Sigmoid Curve)

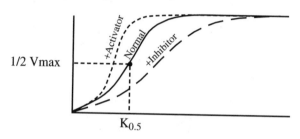

$K_{0.5}$

BIOENERGETICS

ENTHALPY (ΔH)
ΔH is the change in *heat* of reactants and products (i.e., the heat given off with lighting a match).

ENTROPY (ΔS)
ΔS is the change in *randomness* or disorder of the reactants and products.

FREE ENERGY
ΔG = The change in free energy. If ΔG is *negative*, then the reaction occurs spontaneously. If ΔG is positive, then the reaction does NOT occur sponta-neously. $\Delta G° =$ the change in free energy under standard conditions (standard free energy change).

$$\Delta G = \Delta H - T\Delta S$$

$$\Delta G = \Delta G° + RT \ln \frac{[products]}{[reactants]}$$

$$R = 1.987$$
$$T = °K = °C + 273$$

$$\Delta G° = -RT \ln Keq$$

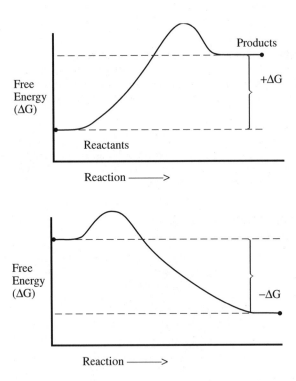

COUPLING

Coupling reactions will cause a reaction to occur that would not normally occur. It will encourage the other reaction to occur by changing the reaction from a $+\Delta G$ to a $-\Delta G$. A large *negative* free energy reaction will "couple" with a smaller positive free energy reaction to yield an overall *negative* reaction. The $\Delta G°$ of ATP is $-7,300$ cal/mol because of the high energy phosphates. Therefore, in the K+/ATPase pump, the overall sum of the reactions will lead to a spontaneous reaction because $\Delta G = +5 + (-7) = -2$.

EXAMPLES OF ENZYMES

Enzyme	Activated by	Inhibited by
Aspartate transcarbamoylase		CTP
Hexokinase (NOT glucokinase)		Glucose-6-phosphate
Phosphofructokinase	Fructose-2,6-Diphosphate, AMP, ADP	ATP, Citrate
Pyruvate kinase		ATP

Enzyme	Activated by	Inhibited by
Protein kinase	Epinephrine	
Phosphorylase b kinase	Glucose-6-phosphate, AMP	Ca^{2+}
Phosphorylase phosphatase		Ca^{2+}
Pyruvate carboxylase	Acetyl CoA	ADP
Pyruvate dehydrogenase	AMP	ATP, NADH, Acetyl CoA
Isocitrate dehydrogenase	ADP	
Fructose-1,6-bisphosphatase	Citrate	AMP
Glycogen synthetase	Glucose-6-Phosphate	
Enzymes requiring *biotin*		Avidin

PYRUVATE DEHYDROGENASE

Involved in oxidative decarboxylation. It uses cofactors, like the α-ketoglutarate dehydrogenase reaction. Pyruvate dehydrogenase reaction joins glycolysis to the TCA cycle. A multienzyme complex is used in this reaction.

PAPAIN

Proteolytic enzyme with esterase, transamidase, and thiol protease activity.

PROENZYMES (INACTIVE ZYMOGENS)

Proenzymes are precursors of enzymes like: trypsin, chymotrypsin, pepsin, and carboxypeptidase (NOT ribonuclease).

PEPSIN

Pepsinogen is released into the stomach, and this proenzyme is activated by acid hydrolysis to form pepsin. It acts as an acid protease or carboxyl protease and is active only in the acidity of the stomach. (Hydrolyzes peptide bonds at low pH.)

OTHER PROTEINS

The majority of proteins in the body are not enzymes, but the structural proteins that make up most of the tissues of the body.

HEMOGLOBIN (HB)

Hemoglobin is a globular protein made up of 4 polypeptide chains; 2 alpha and 2 beta chains. Hemoglobin A subunit structure in adults: $\alpha_2\beta_2$. Each chain contains a heme molecule. A heme molecule contains one iron atom (Fe^{2+} or ferrous state for normal hemoglobin). Each iron can bind 1 oxygen molecule (O_2). Therefore, since there are four chains in a heme molecule, a total of 4 oxygens can be carried by each hemoglobin. The major Hb at birth: HbF, with subunit structure $\alpha_2\gamma_2$. Hemoglobin F binds oxygen *tighter* than HbA does. Hemoglobin can transport CO_2.

Decreased affinity of hemoglobin for oxygen results from: Increased levels of CO_2, increased levels of 2,3-DPG (diphosphoglycerate) in RBC's, acidosis or lowered pH (the Bohr effect), and increased temperature (all shift the oxygen-dissociation curve to the *right*).

Decreased partial pressure of O_2 will decrease the percent of oxygen saturation of hemoglobin. Binding of carbon monoxide and cyanide to Hb is tighter than oxygen binding to Hb. Carbon monoxide inhibits cytochrome oxidase and therefore cellular respiration. Carbon monoxide binds to Hb at the same site as oxygen. Deoxyhemoglobin is stabilized by salt bridges that cross-link poly-peptide chains. Deoxyhemoglobin is a weaker acid than oxyhemoglobin.

Methemoglobin can**not** carry oxygen since the iron is in the ferric state (Fe^{3+}). Small amounts of methemoglobin and carboxyhemoglobin are found in normal people.

Glycosylated Hb: **Hb A$_{1c}$** concentration may be increased in patients with *diabetes mellitus*. The amount of glycosylated hemoglobin depends on the glucose level in the blood; a normal value is between 3 to 9% of the Hb ($\alpha_2\beta_2$-glucose). The Hb A$_{1c}$ concentration reflects the blood glucose level for the previous few months, and is helpful in assessing if the patient is compliant with insulin treatment and diabetes control.

Sickle cell anemia results from substitution of *valine* for *glutamate* at position **6** on the β-chains in hemoglobin (HbS). Valine has a neutral charge, and it replaces glutamate (−1). The sickle shape favors the deoxy form and this leads to a crisis.

HbC is another hemoglobin variant that has a modified β-chain at position **6**, but in HbC, *lysine* (+1 charge) replaces glutamate.

Myoglobin	Hemoglobin
Found in heart and skeletal *muscle*. Is an oxygen carrier	Found in *red blood cells* only. Carries oxygen from the lungs to the tissues, and carries CO_2 from the tissues to the lungs.
Higher oxygen affinity than Hb $P_{50} = 1$ mmHg *Hyperbolic* curve	$P_{50} \approx 26$ mmHg *Sigmoidal* oxygen dissociation curve.
	Cooperative binding, where binding one oxygen increases the affinity to bind.
Monomer; 1 heme group (binds 1 oxygen)	Tetramer; 4 heme groups (HbA = $\alpha_2\beta_2$). (Each subunit has 1 heme, and therefore the 4 heme groups will be able to bind 4 oxygens.)

Myoglobin	Hemoglobin
Binds oxygen more tightly	Binds oxygen less tightly. The ideal oxygen carrier. **T** form = Taut, No oxygen bound, Low oxygen affinity. **R** form = Oxygen is bound, High oxygen affinity.

Oxygen Dissociation Curve

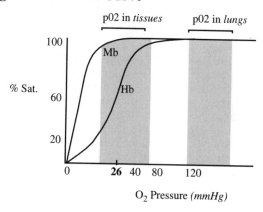

(% Sat. = % of binding that is achieved)

2,3-DIPHOSPHOGLYCERATE (2,3-DPG)

Also known as 2,3-**bis**phosphoglycerate. 2,3-DPG is an intermediate that plays a role in regulating the affinity of hemoglobin for oxygen. At physiologic pH, 2,3-DPG has a net *negative* charge. In red blood cells, it influences the oxygen affinity and binds to **de**oxyhemoglobin (**T** form of Hb)—NOT to oxyhemoglobin (**R** form of Hb). This increase in 2,3-DPG causes a shift to the right in the oxygen-binding curve.

PORPHYRINS

Porphyrins have side chains attached to each of the 4 *pyrrole rings*. An example is **uro**porphyrin III. *Congenital erythropoietic porphyria* can be diagnosed by the pathognomonic type **I** porphyrin in the urine. *Acute intermittent poryphyria* is NOT photosensitive, and results in an increase in δ-ALA and porphobilinogen. The urine darkens in the presence of light or air. Some medications are contraindicated in porphyrias (i.e., barbiturates), since there is an increase in cytochrome p450 and heme to produce cytochrome p450; this decreases the amount of heme even more.

GLYCOSAMINOGLYCANS

Glycosaminoglycans include: hyaluronic acid, heparin, chondroitin sulfate, heparan sulfate, dermatan sulfate, and keratan sulfate (NOT collagen). Glycosaminoglycans are in the extracellular matrix of connective tissue.

PROTEOGLYCANS

Proteoglycans = Protein core + Glycosaminoglycans. Proteoglycans are found in ground substance.

COLLAGEN

Collagen goes through *Hydroxylation* then *glycosylation*. Three pro-α strands join to form a single triple helix **pro**collagen strand. (Gly–X–Y)–(Gly–T–U)–(Gly–. . .) The procollagen strand is transported from the cell. Segments of the N and C terminals of the polypeptides are removed. These procollagen strands spontaneously form the triple helix collagen molecules known as *tropocollagen*. Then, *crosslinks* between the strands form. This occurs between a *lysyl* on one strand and a lysyl on another strand (as well as hydroxylysyl).

Unlike elastin, collagen is *stiff*. Type **I** collagen is found in *bone* and *skin*. The primary structure of procollagen contains many lysines, hydroxylysines, and glycines. *Glycine* allows rotation to occur within the procollagen strands. Collagen is the major protein of connective tissue, tendons, bone, and cartilage. It contains increased glycine, proline, alanine, hydroxyleucine, hydroxyproline, and hydroxylysine residues as a result of *post-transla**tional** modification*. Collagen requires *ascorbic acid* in order to hydroxylate. *Scurvy* results from a deficiency of Vitamin C, and results in increased bleeding, loose teeth in gum and other problems associated with collagen defects.

ELASTIN

Elastin is also found in connective tissue and will stretch (elastic properties) and recoil. Enzymes called *elastases* can break the elastin fibers down. A deficiency in α_1-*antitrypsin* can result in the chronic obstructive pulmonary disease *emphysema*. Normally α_1-antitrypsin inhibits elastase. If there is a gene defect, and the liver cannot produce the α_1-antitrypsin, then the individual may develop this loss of pulmonary function (Non-smoking cause of emphysema).

KERATIN

Keratin is a protein or *intermediate filament* found in epithelial cells (epidermis). It is high in cystine and sulfur (due to increased disulfide amino acids).

IMMUNOGLOBULINS

Immunoglobulins are structural proteins made of pairs of polypeptide chains (light and heavy chains). Immunoglobulins are also classified according to the properties of the *H* chains. Immunoglobulins form the antibodies. Light chains are the κ or λ chains. Heavy chains determine the Class: IgG = γ, IgA = α, IgM = μ, IgD = δ, IgE = ε. (*For more on immunoglobulins, refer to Microbiology & Immunology of this review series.*)

MYOSIN

Myosin is found in skeletal muscle, and myosin fibers are the *thick* filaments (15 nm). Myosin is considered an *ATPase* involved in contraction—it binds to the thin filaments.

ACTIN
Actin is in skeletal muscle, and it is the *thin* filaments (7 nm). There are two forms of actin: the Fibrous (F-actin) form and the Globular form (G-actin).

TROPONIN-C
Troponin-C initiates skeletal muscle *contraction* by binding Calcium.

HOMOGENTISATE
Otherwise known as "Alkapton". This acid is an intermediate of tyrosine breakdown that accumulates in "black urine disease" or alkaptonuria (autosomal recessive disease). It is caused by a deficiency of homogentisate 1,2-dioxygenase. It is part of the degradation pathway of Phenylalanine ➝ Tyrosine ➝ fumarate + acetoacetate (*Lack of homogentisate oxidase*).

IONOPHORES
Ionophores form a complex with a certain ion and can increase membrane permeability to the ion. In some circumstances, antibiotics can act as ionophores. They have a hydro**philic** inside (carboxyl groups).

BLOOD

SYNTHESIS OF HEME
The synthesis of heme occurs in all cells except mature red blood cells (because they lack mitochondria).

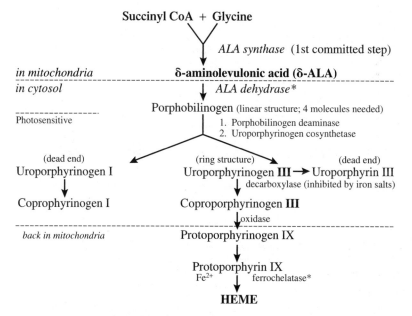

*ALA dehydrase and Ferrochelatase are the two enzymes affected by *lead poisoning*. Therefore, ALA and coproporphyrin increase.

HEME DEGRADATION

Hemoglobin ⟶ Heme + Globin
(The conversion of Heme to Bilirubin is the only reaction in the body that makes carbon monoxide.)

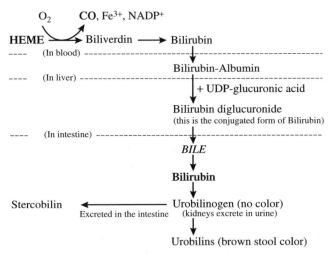

The average life span for a red blood cell is **120 days**.

Unconjugated Bilirubin (Indirect BR)	Conjugated Bilirubin (Direct BR)
Not soluble—aqueous	+ Solubility (water soluble)
+ Lipid solvent solubility	NOT lipid soluble
Attaches to plasma albumin	NO covalent attachment
In Brain	In Icteric urine and bile
Hemolytic jaundice is greater	Intrahepatic and posthepatic jaundice is greater

Stercobilin
This is the degraded product of hemoglobin that creates the brown color of feces.

Type O Blood individuals (Universal donors)
Type O Blood individuals have the H specificity on glycoproteins in secretions. They have the H gene and produce fucosyl transferase. They have anti-A **and** anti-B antibodies in the serum—they do **NOT** have A or B **antigens**.

Type A Blood individuals
Have anti-B antibodies in the serum. They have the **A antigen**. They cannot receive a blood transfusion from type B blood. The A antigen specificity may

Blood Clotting

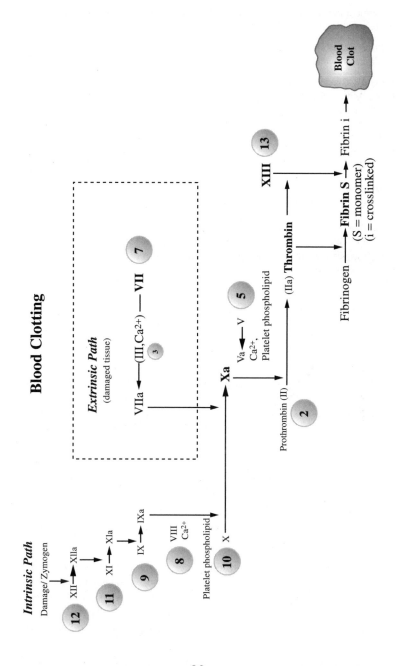

be determined by the terminal monosaccharide on the oligosaccharide of glyco-proteins and glycolipids.

Type B Blood individuals
Have anti-A antibodies in the serum. They have the **B antigen**.

Type AB Blood individuals
Have **A and B antigens** on glycoproteins in body fluids. They also synthesize the H antigen.

Bleeding can be caused by:
Decreased thromboxane A_2, Vitamin K, Factor VIII or antihemophilia A factor, and α_2-antiplasmin.

Thrombosis (clotting) is *prevented by*:
Prostacyclin, Heparin, Dicoumarol, Antithrombin III (NOT antiplasmin).

Thrombin
Enzyme formed from prothrombin that *causes blood clotting*. Removes fibri-nopeptides from fibrinogen. Converts **fibrin**ogen to fibrin (NOT plasminogen to plasmin).

Plasmin
An enzyme with proteo**lytic** action that *dissolves* a clot. Acts as a fibrinase.

Summary of Protein Separation Techniques	
Affinity chromatography	Used to purify *specific* proteins. *Covalent binding* of protein (Ab, receptor, substrate) to column material.
Separates proteins by **charge**	
Ion-exchange chromatography *Electrophoresis*	Net *charge*, *migration* in electric field.
Separates proteins by **size**	
Gel-filtration chromatography	Large molecules move more quickly through column (first proteins to come out), smaller go between beads.
Dialysis	Separates by *size* (large from small)

DNA, RNA, and Protein Synthesis

4

DNA AND RNA

REPLICATION   DNA ⟶ DNA

TRANSCRIPTION   DNA ⟶ RNA (mRNA, tRNA, rRNA)

TRANSLATION   mRNA (codons)

           ⟵ tRNA (anti-codons) + activated amino acids

    rRNA⟶

    Proteins (structural, enzymes, globins, contractile, etc.)

Remember, you "translate" nucleic acids into proteins.

NITROGEN BASES

Pyrimidines: ("CUT")
    Cytosine (C), Uracil (U), Thymine (T)
Purines:
    Adenine (A), Guanine (G)

Note that the basic structure of the purines is an imidazole ring joined at carbons 5 and 6 of the pyrimidine ring. The sum of purines is equal to the pyrimidines.

PENTOSE (5 CARBON) SUGARS

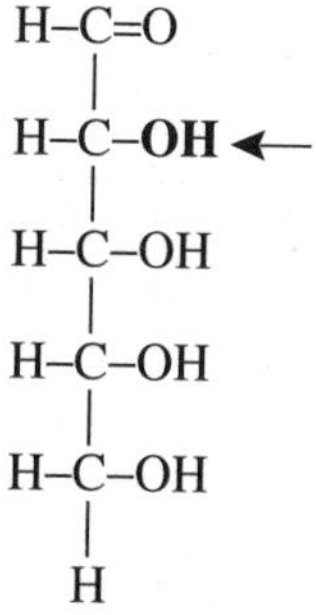

$$
\begin{array}{cc}
\text{H–C=O} & \text{H–C=O} \\
| & | \\
\textbf{H–C–OH} \leftarrow & \textbf{H–C–H} \leftarrow \\
| & | \\
\text{H–C–OH} & \text{H–C–OH} \\
| & | \\
\text{H–C–OH} & \text{H–C–OH} \\
| & | \\
\text{H–C–OH} & \text{H–C–OH} \\
| & | \\
\text{H} & \text{H} \\
\textbf{Ribose} & \textbf{Deoxyribose}
\end{array}
$$

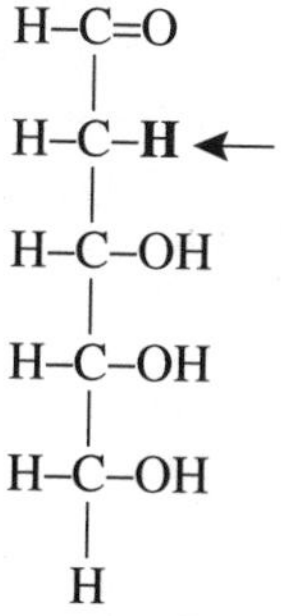

35

PHOSPHATE

$$
\begin{array}{c}
\text{O} \\
\| \\
\text{O}=\text{P}-\text{O} \\
| \\
\text{O}-
\end{array}
$$

NUCLEOSIDE

A nucleoside is a nitrogen **base** and a pentose **sugar**. Most anti-cancer drugs are nucleoside analogs.

NUCLEOTIDE

A nucleotide is a **nucleoside** plus **phosphate** joined at pentose sugar; nucleotides are the building blocks of RNA and DNA. Also, nucleotides serve as the medium of energy exchange inside the cell, most commonly in the form of ATP.

ATP (ADENOSINE TRIPHOSPHATE)

A ribose nucleotide which stores energy from catabolic metabolism in its high energy phosphate bonds.

DNA (DEOXYRIBONUCLEIC ACID)

Polynucleotide chain formed by joining together **deoxy**ribose nucleotides with phosphodiester bonds between the 5'-hydroxyl and the 3'-hydroxyl groups on adjacent deoxyriboses. By convention, the sequence of DNA is written from the free 5' end of the molecule to the free 3'-hydroxyl end. DNA encodes instructions which comprise the genetic code. Ribose is converted to deoxyribose by the enzyme *ribonucleotide reductase*. This enzyme insures that the amount of Adenine will be equivalent to Thymine, and Guanine equivalent to Cytosine.

RNA (RIBONUCLEIC ACID)

Contains ribose nucleotides and the nitrogen base *Uracil* instead of Thymine (see below). It is involved with carrying out the instructions encoded for by DNA. The pentose phosphate pathway makes ribose. The 2' Hydroxyl group can be ionized; therefore, RNA is less stable than the DNA.

DNA AND RNA PURINES = Adenine, Guanine

DNA PYRIMIDINES = Cytosine, Thymine

RNA PYRIMIDINES = Cytosine, Uracil

BASE PAIRS

Nitrogen bases form complementary base pairs by hydrogen bonding between one purine and one pyrimidine. The sum of the purines is equal to the pyrimidines. Charkof's Rule is A=T and G=C, so as long as you know one of the values, you can calculate the other three. In humans, 30% is A (30% is therefore T) and 20% is G (20% is therefore C), for a total of 100% of bases.

- Guanine only base pairs with cytosine (3 H-bonds; G=C)
- Adenine base pairs with a thymine from DNA, or an uracil from RNA (2 H-bonds; A=T, A=U)
- Base pairing can hold 2 different polynucleotides together

Purines	Number of H-bonds	Pyrimidines
Guanine	3	Cytosine
Adenine	2	Thymine in DNA
	2	or Uracil in RNA

MELTING TEMPERATURE

The more G–C bonds between the two strands of dsDNA, the more they are stabilized. This will increase the Tm (melting temperature) at which the two strands will separate.

NUCLEASES

Nucleases are enzymes which hydrolyze phosphodiester bonds

Exonuclease

Cleaves nucleotides from the *end* of a polynucleotide chain

Endonuclease

Break polynucleotide by cleaving at nucleotides *inside* of the chain.

Restriction Endonucleases

Enzymes, produced by bacteria and viruses, which recognize and cleave at short specific sequences of a polynucleotide called *palindromes*. These enzymes are used in genetic engineering. They are bacterial defense mechanisms used to fend off an invading DNA. They protect by methylation.

PALINDROME

A short series of nucleotides which has the same sequence as its complementary strand and is read from **5' to 3'**. It can be read the same in either direction (i.e., **5'**–GAATTC–3' ←→ 3'–CTTAAG–**5'**). Notice that in the example, it reads the same in the 5' to 3' direction, GAATTC. An EcoR I restriction site can allow cleaveage and re-alignment with other genes to clone and express genes.

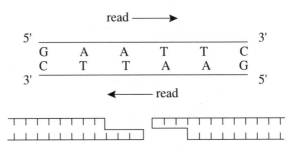

EcoR I restriction site

DNA STRUCTURE

- Double Helix, except in some viruses (Parvoviruses have **ss**DNA).
- 2 antiparallel polynucleotide chains one strand running 5'–3' and the other 3'–5', held together by complementary base pairing.
- Contained in *cytoplasm* of **pro**karyotes (before the nucleus).
- Contained in the *nucleus* of **eu**karyotes (true nucleus).
- The middle of the helix is held together by *hydrogen* bonds.
- Is mildly acidic.
- Mitochondria have their own DNA.
- **3'** to 5' *phosphodiester bonds* hook the nucleotides together.
- Most proteins bind in the *major* grooves (not the minor grooves).

3 TYPES OF DOUBLE HELIX

B form: right-handed helix with 10.5 base pairs per turn.

A form: right-handed helix with 11 base pairs per turn and is formed by dehydrating B form.

Z form: *left*-handed helix with 12 base pairs per turn occurs in small sections of DNA. It has an alternating anti- and syn-glycosidic bond conformation.

PROKARYOTES

Prokaryote means "before the nucleus;" it includes bacteria and blue-green algae.

PROKARYOTIC DNA

One large circular DNA chromosome located in the *cytoplasm*—since there is no nuclear membrane to hold on to.

PLASMIDS

Small *circular DNA* molecules also found in the *cytoplasm* of some bacteria code for extrachromosomal genes. Replication and inheritance of these genes is independent from that of the chromosome. More than one type of plasmid may be present in a bacteria and a single bacteria can have hundreds of plasmids. Plasmid genes have often been found to carry genes encoding for antibiotic resistance. Under certain circumstances plasmid DNA is incorporated into the large circular chromosome and its DNA is then replicated and inherited with it.

MITOCHONDRIAL DNA

Circular DNA chromosome similar to the large bacterial chromosome.

EUKARYOTES

Eukaryote means "true nucleus." Eukaryotes are highly organized organisms; plants, animals and single-celled organisms, except bacteria and blue-green algae. Their genetic material is contained within an intracellular nuclear membrane.

EUKARYOTIC DNA

DNA organized into chromosomes and located inside of the nucleus.

CHROMOSOME

Prokaryotes have **1** large chromosome which replicates and divides at the time of cell division. Eukaryotes have much more DNA, which is divided into a number of chromosomes. Every organism has a specific number of chromosomes (46 in humans). Each chromosome is replicated and then evenly divided during cell division assuring that each cell gets the right number of chromosomes. Humans are *diploid* meaning that there are 2 of each of the 23 types of chromosomes (one inherited from the father and 1 from the mother).

PACKING OF DNA

NUCLEOSOMES

DNA is organized into clumps called *nucleosomes* by complexing with histones, giving it the appearance of "threaded beads." DNA wraps around the histone 2 times. A nucleosome is a DNA-wrapped histone core! DNA wraps around histone core

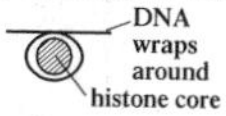

—o—O—o—O—o—O—o—O—o—O—o—O—o—O—o—O—o—

HISTONES

Small arginine and lysine rich basic proteins:

—o— H1 1 molecule complexed with thread portion of DNA.
—O— Histone core is formed from 8 histone molecules around which ~140 base pairs are wound comprised of:
   H2A 2 molecules
   H2B 2 molecules
   H3 2 molecules
   H4 2 molecules

CHROMATIN

Further packing of DNA due to hydrophobic interactions and in association with other non-histone proteins compacts it into chromatin.

Heterochromatin
This is very densely packed and inactive chromatin.

Euchromatin
This is **active** chromatin.

REPLICATION OF DNA

DNA ———▶ DNA

Replication is by semiconservative replication:

After replication each of the new double helixes is made of one original strand and one newly synthesized strand.

1. Double stranded DNA unwinds and 2 single strands are exposed forming a "replication fork" which moves along the chain as DNA is replicated. This occurs in *both* directions with 2 replication forks in an anti-parallel, 5' to 3' direction. These processes are facilitated by the following enzymes:

   ### DNA Helicase
   Binds to single stranded DNA at the replication fork and opens the double stranded DNA similar to pushing a zipper apart. This unwinds the DNA, and it requires ATP to open the DNA.

   ### Helix-destabilizing proteins
   Binds to and stabilizes single strands of DNA. The helix-destabilizing proteins are also called ssDNA-binding proteins. As the replication fork proceeds it causes twisting which is relieved by topoisomerases.

   ### Topoisomerases
   Type I (swivelase) binds to and cleaves *one* of the single strands of DNA (nuclease activity), allows the DNA to untwist around the axis of the phosphodiester bond of the intact DNA strand and then reconnects the single strand (ligase activity).
   Type II (gyrase) binds to and cleaves *both* single strands at the same time and results in negative twisting which relaxes both strands and then reseals the strands. Quinolones inhibit DNA gyrase.

   ### Ligases
   Connect two strands of DNA end to end. This enzyme seals the DNA strands. The process requires ATP or NAD+.

2. **Primase** (an **RNA** polymerase) uses triphosphate ribonucleotides to form a short strand of RNA complementary to DNA near the replication fork which serves as the double stranded *primer* necessary for DNA polymerase

III. Later, **DNA** polymerase I (**exo**nuclease) removes the RNA primer and replaces it with DNA.

3. **DNA polymerase III** *reads* each of the old strands in the 3' to 5' direction and uses the appropriate triphosphate deoxyribonucleotides to synthesize new complementary strands in the **5' to 3'** direction. The energy to drive these reactions is provided by the high energy triphosphate groups; these are cleaved and form the monophosphates that are incorporated into the DNA chain. DNA polymerase III also *proofreads* the newly synthesized strands removing any incorrect nucleotide and inserting the correct one. This enzyme is supposed to correct the 1/10,000 mutations occurring during DNA replication. The 3'–5' exonuclease activity (proofreading) allows for checking the new chain as it grows in the 5'–3' direction.

4. The new strand being synthesized complementary to the old 3' to 5' strand is called the **leading strand** and can be synthesized without interruption (it is copied continuously).

5. The new strand complementary to the old 5' to 3' strand is called the **lagging strand** and is synthesized as fragments called **Okazaki fragments**, which are later joined together by ligases. The lagging strand is copied "**dis**continuously".

DNA polymerase I	DNA polymerase II	DNA polymerase III
5' to 3' elongation	5' to 3' elongation	5' to 3' elongation
3' to 5' exonuclease	3' to 5' exonuclease	3' to 5' exonuclease
5' to 3' exonuclease		
Prokaryotes = I	Prokaryotes = II	Prokaryotes = III
Eukaryotes = α	Eukaryotes = β	Eukaryotes = γ

PROKARYOTES
Have 1 site of initiation of replication called the "origin of replication," or *replicon*. In DNA repair or excision, DNA polymerase **I** recognizes the gap (nick), removes the RNA primer, and fills the gap. DNA polymerase **III** is responsible for synthesizing the new DNA.

Types of polymerase	Function
DNA Polymerase I	DNA repair
DNA Polymerase II	Chain elongation
DNA Polymerase III	DNA replication

EUKARYOTES
Have many sites at which replication of DNA may be initiated.

Types of polymerase	Function
DNA Polymerase α	DNA replication
DNA Polymerase β	DNA repair
DNA Polymerase γ	(replication in mitochondria)

Template sequence 5'–ApCpTpGpGp–3' would make the complementary structure: 5'–CpCpApGpTp–3'. This is determined by forming C for G, and A for T; then, always write 5' → 3' (using 3' → 5' original sequence). *If this is RNA replication, then place a "U" for any "A."*

DNA **synthesis** *always* occurs in the unidirectional **5' to 3'** direction. DNA synthesis *always* requires a *primer*—RNA polymerase or primase is required (RNA synthesis does not require a primer).

DNA REPAIR

1. RADIATION

U.V. endonuclease
Removes pyrimidine-pyrimidine dimers that occur, usually between two thymines (T-T), in the same strand due to exposure to ultraviolet light. Then polymerase I or α fills in the gap with the appropriate nucleotides. Dimers can lead to mutations. Therefore, finding the mistake and repairing it is important.

Xeroderma Pigmentosum
Disease occurring in people who cannot repair damage caused by U.V. light, often due to a defect or deficiency in U.V. endonuclease. This leads to skin cancer—malignant melanoma.

2. DEAMINATION
Loss of an amino group.

Alkylating Agents
Remove $-NH_3$ groups from A, G, or C, forming mutant bases.

Nitrous Acid
Alkylating agent produced in cells by metabolism of nitrites and nitrates which are often used as food preservatives.

Cytosine
Can spontaneously lose its $-NH_3$ group forming uracil.

Mutant Base Endonucleases
A group of enzymes each of which recognize and remove a specific mutant base containing nucleotides.

Mutant Base Glycosylases
Enzymes which recognize and remove specific mutant nitrogen bases from their deoxyribose.

DNA Transcription to RNA (RNA Synthesis)
DNA ⟶ RNA

RNA
Here are some generalizations for RNA: RNA is single-stranded except in some viruses (i.e., Rotavirus or Reovirus). It contains ribose instead of deoxyribose, and *uracil* instead of thymine. Base pairing may occur within the single strand. Base pairing can occur between RNA and DNA. There are 3 types of RNA involved in protein synthesis: **mRNA**, **tRNA**, and **rRNA**.

PROKARYOTIC DNA TRANSCRIPTION INTO RNA
Takes place in the *cytoplasm*.

I. INITIATION
A short area of single stranded DNA is exposed by the binding of RNA polymerase at a site known as the *promoter region*. A holoenzyme (σ-factor complex) recognizes this specific DNA binding site. The promoter is the region of DNA that binds the RNA polymerase to initiate transcription. Usually the first base is a purine (Adenine).

II. ELONGATION
As RNA polymerase moves along, it maintains an area of single stranded DNA on either side of it. **R**NA polymerase *reads* the DNA strand in the **3'** to **5'** direction and uses triphosphate ribonucleotides to *synthesize* an RNA strand in the **5'** to **3'** direction as in DNA replication (always *grows* in the **5'** to **3'** direction). There is NO proofreading in RNA polymerase since there is NO RNA 3' to 5' exonuclease activity—this results in more errors when transcribing DNA to RNA (compared to the relatively error-free DNA to DNA replication).

III. TERMINATION
1. Rho-independent termination. In prokaryotes, termination of RNA synthesis is coded for by *palindrome* DNA sequences. RNA transcribed from these sequences can base pair within the sequence forming a *hairpin loop* causing termination.
2. ρ (Rho) factor destabilizes RNA and DNA, and may help in the termination of transcription.

    Note: transcribed is synonymous with transcripted.

PROKARYOTES
Promoter allows RNA polymerase to transcribe DNA, and it contains 2 specific consensus sequences:

1. TATAAT known as Pribnow box, ~10 bases before the first transcripted DNA sequences.
2. TGTTG ~35 bases before the first transcripted DNA sequences.

Promoter
5'- / / -- TGTTG ------- TATAAT ------ Transcripted DNA --/ / -3'
|◄——— ~ 35 bases ———►|
|◄—►|
~ 10 bases

CONSENSUS SEQUENCE
A DNA sequence which is representative of many promoters from different bacteria.

PROKARYOTIC RNA POLYMERASE
There is only **one** RNA polymerase in **pro**karyotes. This is a DNA-directed RNA polymerase. A holoenzyme is composed of 5 subunits: $\alpha\alpha\beta\beta'\sigma$ [two of the α unit, one each of the β and β' units ($\alpha\alpha\beta\beta'$ = core enzyme), plus a cofactor called σ (sigma), which is required for the recognition of the promoter during initiation]. *Sigma factor* (a subunit of RNA polymerase) gives the specificity of initiating transcription of RNA. In bacteria, both the sigma factor and the core enzyme of RNA polymerase are required for RNA synthesis.

TYPES OF PROKARYOTIC RNA

I. mRNA
Contains the codons (the 3 nucleotide sequences) which specify each of the amino acids in a polypeptide chain. mRNA may be *translated* into a protein without further processing. Translation can begin at the free 5' end before the mRNA is completely transcribed. Prokaryotic mRNA does not contain a "cap" on its 5' end—eukaryotic mRNA does. mRNA contains many "cistrons" (polycistronic) one after the other, each cistron coding for one protein. Many proteins may be translated from the same mRNA at the same time, one after the other. It has a short half-life.

II. tRNA
Contains the anti-codons which are complementary to the codons of the mRNA. Codons and anti-codons base pair during translation (protein synthesis). Transports activated amino acids; is cleaved from a larger precursor. D-loop and TψC loop are areas that do not base pair. The cloverleaf structure of tRNA is short, has 70 to 90 bases, and many *unusual bases*.

III. rRNA
A 30S precursor rRNA molecule is cleaved to form one of each; a 5S rRNA, a 16S rRNA, and a 23S rRNA. Ribosomal RNA.

S (Svedberg unit)
A measure of the sedimentation rate of a molecule.

PROKARYOTIC RIBOSOMES

1. The small 30S ribosomal subunit is made from two 16S rRNAs.
2. The large 50S subunit is made from 5S and 23S rRNAs.
3. The entire 70S ribosome contains one 30S subunit and one 50S subunit.

EUKARYOTIC *DNA TRANSCRIPTION INTO RNA*

- Takes place in the *nucleus* and then the finished RNAs (except for small nuclear RNAs) are transported into the cytoplasm, where protein synthesis takes place on the ribosomes.
- Similar to transcription in prokaryotes with the following exceptions:

EUKARYOTIC PROMOTERS

1. TATATAA—known as Hogness box or TATA box ~25 bases prior to transcribed DNA.
2. GGTCAATCT—known as CAAT box ~70 bases before transcripted DNA.
3. GC rich areas (GC boxes) occurring approximately 40 to 110 bases before the transcribed DNA.

```
              Promoter region
 ┌──────────────────────────────────────────────┐
5'- / / --[GCGC--]--GGTCAATCT----TATATAA----- transcripted DNA--/ / - 3'
          |◄────── ~ 40-110 bases ──────►|
                   |◄── ~ 70 bases ──►|
                              |◄►|
                           ~ 25 bases
```

TRANSCRIPTION FACTOR (TF)

TF is similar to the prokaryotic sigma factor. Transcription factor binds to DNA in the major groove.

EUKARYOTIC RNA POLYMERASE

There are 3 types of RNA polymerase
RNA polymerase **I**: synthesizes a 45S precursor **r**RNA which is cleaved to form: a 5.8S, a 18S, and a 28S rRNA.
RNA polymerase **II**: synthesizes **m**RNA.
RNA polymerase **III**: synthesizes **t**RNA and the **5S r**RNA.

EUKARYOTIC RNA

Eukaryotic mRNA is always **mono**cistronic. RNA is single stranded (ss), it has a *negative* charge at neutral pH, and it has **R**ibose sugar (not **d**eoxyribose **D**NA). Modify eukaryotic mRNA after transcription with 3'-polyadenylate tails. The 5'-Caps only occur in **eu**karyotic **m**RNA.

TYPES OF EUKARYOTIC RNA

I. rRNA

45S precursor synthesized by RNA polymerase I is cleaved to form the 5.8S, 18S, and 28S rRNA chains. RNA polymerase III synthesizes the 5S rRNA chain. Approximately 80% of RNA in cells is rRNA. It is found in ribosomes and the nucle**olus**.

II. mRNA

- Contains *codons* (the 3 nucleotide sequences which specify the individual amino acids).
- The RNA chain synthesized by RNA polymerase II is called heterogenous RNA (hnRNA).
- Some hnRNAs contain intervening RNA sequences called "*introns*" which do not code for amino acids and which must be removed from the chain *before* translation can occur, and *before* mRNA is transported out of the nucleus (post-*transcriptional* modification).
- "*Exons*" which contain the codons that are translated into proteins are spliced together in a process which uses small nuclear RNAs (snRNA). snRNA are small nuclear RNAs, and are found in spliceosomes. snRNP, "snurp," is **snRNA** and *Protein* (ribonucleo-protein complex = **RNP**).
- Poly (A) polymerase synthesizes a tail connected to the 3' end made up of many adenosine nucleotides of some mRNAs in the nucleus; after transport into the cytoplasm the tail is removed.
- Guanosine triphosphate is attached to the 5' end and then gets a methyl $-CH_3$ group on carbon number 7, forming a "Cap" on that end.
- Post-*transcriptional* modification occurs *only* with **eu**karyotic mRNA:
  1. Addition of 5'-7-met-Guanosine cap helps in translation—this is a rare 5' to 5' linkage.
  2. Addition of 3' Poly-A tail for stabilization.
  3. Removal of introns or *intervening sequences.*

III. tRNA
Contains anti-codons complementary to mRNA codons. A *CCA* sequence is added to the 3' end. Contains *unusual* nitrogen bases—not analogs (N-acetyl-cytosine, pseudouracil, and dimethyladenine). Made from larger precursor.

EUKARYOTIC RIBOSOMES
- The small 40S ribosomal subunit is formed from two 18S RNAs.
- The large 60S ribosomal subunit is formed from proteins and the 5S, 5.8S, and 28S rRNAs.
- The entire **80**S ribosome is formed from one 40S and one 60S ribosomal subunit.

REVERSE TRANSCRIPTION
Synthesis of DNA from RNA using viral "RNA directed DNA polymerase."

POST-TRANSCRIPTIONAL MODIFICATION
Again, this is modification of RNA. It can occur by the addition of terminal sequences; for example: at the 3'end, addition of 5'-CCA-3' to the tRNA. Also,

methylation of bases (*met*-Guanine) and methylation of sugars (2'-OH group can be methylated). Furthermore, cleaving a ribonucleotide or removing introns.

TRANSLATION Polypeptide (protein) Synthesis

RNA ⟶ PROTEIN

Polypeptides are synthesized from their amino to their carboxyl end.

PROKARYOTIC PROTEIN SYNTHESIS

I. Activation
Each type of amino acid is activated by joining to ATP using a specific aminoacyl-tRNA synthetase (Activation with **ATP**).

The specific aminoacyl-tRNA synthetase is also responsible for binding each type of amino acid to its correct tRNA.

The amino acid-monophosphate-tRNA is called a *charged tRNA*.

The energy to drive these reactions comes from ATP being cleaved to AMP and 2 inorganic phosphates (Pi + Pi).

II. Initiation
An **70S** ribosome complex is made of 30S and 50S *ribosomal subunits*; the ribosome contains amino (A) and peptide (P) binding sites. (Note: Streptomycin inhibits *initiation*.)

IF-3 binds to the 30S ribosomal subunit and causes it to dissociate from the 50S subunit. (IF = initiation factor)

Then, mRNA and IF-1 bind to the small 30S ribosomal subunit.

The 5' end of each mRNA has a *Shine-Dalgarno sequence* (5'–AGGAGGA–3'), about 7 bases before the first codon. (Not a 5'-Guanosine cap like eukaryotes.)

The 16S rRNA of the small ribosomal subunit has a sequence complementary to the Shine-Dalgarno sequence which binds to it.

The ribosome looks for the first AUG mRNA codon that it comes to. This first mRNA codon (**AUG**) always codes for the amino acid *methionine*.

Methionine binds to a specific initiator tRNA (i-tRNA).

After binding to i-tRNA methionine receives a formyl group.

N10-formyltetrahydrofolate is the formyl group donor in a reaction catalyzed by methionine transformylase.

IF-2 binds to the i-tRNA-*formyl* methionine and they bind to the 30S subunit.

As the large 50S ribosomal subunit re-associates with the small 30S subunit forming a complete ribosome and the initiation factors are released.

Release of the initiation factors uses energy from **GTP** (guanosine triphosphate) being cleaved to GDP + Pi.

The **initiation complex** has now been assembled. It includes:

The complete ribosome (30S and 50S ribosomal subunits) bound to mRNA at the Shine-Dalgarno sequence.

The initiator-tRNA-formylmethionine, with its anticodon (CAU) base paired to the mRNA's codon (AUG), held in the ribosome's peptide (P) site with the ribosome's amino (A) site is vacant.

- **Eukaryotes** have several eukaryotic initiation factors (**eIFs**).

- The initial formylmethionine amino acid can later be removed from the polypeptide.

III. Elongation
The mRNA at the A site contains the next codon.

When another charged tRNA-phospho amino acid moves into the vacant A site, its anticodon base pairs with the mRNA codon and it also binds to the ribosome's A site. This process consumes GTP and requires the elongation factor, EF-Tu.

Peptidyl transferase, an enzyme that is part of the 50S subunit, catalyzes a reaction that forms a peptide bond between the *carboxyl* group on the amino acid at the P site and the *amino* group of the amino acid at the A site. (Puromycin and Chloramphenicol inhibit elongation at this step.)

The polypeptide grows from the amino side (**N**-terminal) **to** the carboxyl side (**C**-terminal).

Formation of the peptide bond is driven by the hydrolysis of the phosphate bond from the P site amino acid. This results in an uncharged P site tRNA no longer bound to an amino acid, and the A site tRNA being bound to 2 amino acids (a dipeptide).

And the whole process above may be repeated.

Eukaryotes also have termination factors (eTF) that will end synthesis.

(Tetracyclines prevent elongation.)

IV. Translocation
(translocate tRNA-peptide complex from the A site to the P site)

The ribosome moves 3 nucleotides down toward the mRNA's 3' end.

The uncharged tRNA is released.

The dipeptide tRNA moves to the P site.

A new tRNA phospho amino acid binds to the A site consuming another molecule of **GTP** and requiring EF-Tu, and elongation is repeated. (use 2 GTP's/ peptide bond)

Diphtheria toxin alters EF-2 and stops elongation—can't translocate.

(Clindamycin and Erythromycin inhibit translocation.)

V. Termination
When one of the mRNA stop codons reaches the A site, releasing factors (RF) cause the mRNA and the new polypeptide to be released. (Stop codons or non-sense codons = UAA, UAG, and UGA.)

RF-1 recognizes UAA or UAG.
RF-2 recognizes UAA or UGA.

GENERAL CONCEPTS IN PROTEIN SYNTHESIS:
(mRNA ——▶ Proteins)

General Sequence:
1. Activate amino acids,
2. mRNA, initiation factors, join with ribosomal subunits,
3. GTP, elongation factors, join with aminoacyl-tRNA,
4. Peptide bond forms,
5. Peptidyl-tRNA is translocated.

Use high energy phosphate bonds when:
1. Activate the amino acid,
2. Bind N-formylmethionyl-tRNA to the initiation codon of mRNA (joining of ribosomal subunits),
3. Bind aminoacyl-tRNA to active site of ribosome,
4. Translocation occurs.

POLYSOME
Many ribosomes may be attached to a single mRNA, all of them synthesizing new polypeptides. The entire complex is called a *polysome*.

Polysome

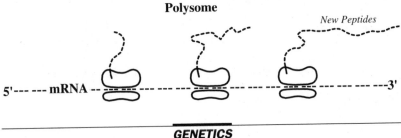

New Peptides

5'--- --- mRNA ---   ---3'

GENETICS

OPERON

DNA of the regulator, promoter, operator, and structural genes.

Regulator

Genes for repressor protein synthesis. *Repressor protein* will bind to the operator, preventing RNA polymerase from binding to the promoter region; inhibits synthesis. (Prevents transcription of structural genes.) A *positive* regulator enhances production of products (CAP = Catabolite gene activator, located upstream from the promoter).

Inducer

Binds to the repressor; *induces* or allows synthesis of RNA.

Lactose

Lactose activates the lac operon to allow formation of: β-galactosidase (y gene), galactoside permease (y gene), and thiogalactoside transacetylase (a gene). (Glucose *decreases* cAMP and inhibits the lac operon. Lactose increases cAMP which activates the lac operon).

The enzymes produced will break down lactose.

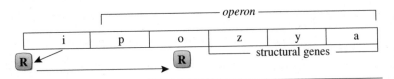

WOBBLE

The nucleotide in the 3rd position has less influence on which amino acid a codon specifies. For example, the codons CUU, CUC, CUA, and CUG all code for the amino acid leucine; this is called *wobble*.

Notice that while each 3 nucleotide sequence codes for only one kind of amino acid, i.e., CUA always codes for leucine, other codons can still code for that amino acid.

The genetic code is called *degenerate*, since more than one codon can specify a given amino acid. It usually does not matter what base is in the third position.

Triplet Code

5'-End ———————————		mRNA ——————————			3'-End
Position 1		Position 2			Position 3
	U	C	A	G	
	Phe	Ser	Tyr	Cys	U
U	Phe	Ser	Tyr	Cys	C
	Leu	Ser	**Stop**	**Stop**	A
	Leu	Ser	**Stop**	Trp	G
	Leu	Pro	His	Arg	U
C	Leu	Pro	His	Arg	C
	Leu	Pro	Gln	Arg	A
	Leu	Pro	Gln	Arg	G
	Ile	Thr	Asn	Ser	U
A	Ile	Thr	Asn	Ser	C
	Ile	Thr	Lys	Arg	A
	Met	Thr	Lys	Arg	G
	Val	Ala	Asp	Gly	U
G	Val	Ala	Asp	Gly	C
	Val	Ala	Glu	Gly	A
	Val	Ala	Glu	Gly	G

The genetic code is represented by a 3 nucleotide base sequence in mRNA molecules. The mRNA molecules read from the 5' to the 3' end; this short sequence is called a *codon*.

The genetic code is universal, it is the same in all organisms—except in ciliated protozoa and in mitochondria.

CONSTITUTIVE GENE
This is a gene that *continually* forms a product.

REGULATED GENE
This is a gene expression based on environmental or developmental factors. For example, *galactosidase* increases when there is no glucose, but there is galactose to break down.

RESTRICTION ENDONUCLEASES
This allows the mapping of the genotype used in *cloning*.

RESTRICTION FRAGMENT LENGTH POLYMORPHISM (RFLP)
This shows the genetic difference between offspring. It is used for many legal custody battles. Phenotypically we may appear identical, but *genetically* there is a difference.

MUTATIONS
Are the result of changes in the **mRNA** *codon* nucleotide sequence.

FRAME SHIFT MUTATIONS
Shift the reading frame of the entire DNA chain after the mutation, include the following:

Deletion mutation
The loss of a single base either spontaneously or due to damage.

Insertion mutation
Acridine intercalates between adjacent DNA nitrogen bases and gets read by RNA polymerase causing the addition of extra bases into the new mRNA.

POINT MUTATIONS
Include the following mutations where only 1 nucleotide is changed:

Transition mutation
When a purine is substituted by a purine.
When a pyrimidine is substituted by a pyrimidine.

Transversion mutation
Substitution of a purine for a *pyrimidine*.
Substitution of a pyrimidine for a *purine*.

Silent mutation, i.e., UUU (Phe) ⟶ UUC (Phe)
Change codon to another codon for the *same* amino acid, so has no effect.

Nonsense mutation, i.e., UGG (Try) ⟶ UGA (stop)
Change codon to a *stop codon* (**UGA, UAG,** and **UAA**) and terminates synthesis.

Missense mutation, i.e., UUC (Phe) ⟶ UUA (Leu)
Changes codon to another codon for a *different* amino acid. If the new amino acid is similar to the old one the synthesized protein might function.

ANTIBIOTICS BY SITE OF ACTION

Many antibiotics are selectively toxic to either eukaryotic or prokaryotic organisms because they interfere with specific enzymes of DNA, RNA, or protein synthesis.

A. PROTEIN SYNTHESIS INHIBITORS

30S ribosomal subunit

Aminoglycosides (i.e., streptomycin) bind to prokaryotic 30S ribosomal subunit and causes misreading (prevents the initiation of protein synthesis). *Tetracycline* binds to the acceptor site on the prokaryotic 30S subunit preventing it from binding the activated amino acid-tRNA complex.

50S ribosomal subunit

Chloramphenicol binds to prokaryotic 50S subunit and inhibits peptidyl-transferase.
Erythromycin and *Clindamycin*:
Both bind to the 23S rRNA within the prokaryotic 50S subunit and prevent translocation.

60S ribosomal subunit

Cycloheximide binds to the eukaryotic 60S subunit and inhibits peptidyltransferase.

A and P sites

Puromycin is an amino acid analog that binds to the A site and has an amino group which can form a peptide bond. No further elongation can take place, and protein synthesis of both prokaryotes and eukaryotes is inhibited.

eEF-2

Diphtheria toxin inhibits eukaryotic elongation factor 2; interrupting eukaryotic protein synthesis.

B. RNA SYNTHESIS INHIBITORS

Prokaryotic DNA-directed RNA polymerase

Actinomycin: binds to double stranded DNA so that RNA polymerase cannot read it.
Rifampin: binds to the β-subunit of RNA polymerase and inhibits the start of transcription.

Eukaryotic RNA polymerase II

Amanita phylloides—the "Angel of Death" mushroom produces a toxin which inhibits RNA polymerase II stopping **mRNA** production.

C. DNA SYNTHESIS INHIBITORS

DNA gyrase inhibitors (DNA synthesis inhibitors)
Quinolones bind to and inhibit DNA gyrase preventing DNA synthesis.

DNA topoisomerase inhibitors
Nalidixic acid and *Novobiocin*:
Both inhibit DNA topoisomerase, stopping DNA synthesis.

Drug	Mechanism of action
Aminoglycosides (i.e., Streptomycin)	Inhibits protein synthesis (translation) by binding to the **30S** ribosomal subunit.
Tetracycline	Inhibits translation by binding the **30S** ribosomal subunit.
Chloramphenicol	Inhibits translation by binding to the **50S** subunit and inhibits peptidyl-transferase.
Erythromycin Clindamycin	Binds to the 23S rRNA in **50S** subunit (inhibits translation in **pro**karyotes)
Cycloheximide	Binds to the **60S** subunit, inhibits peptidyltransferase, prevents translation in **eu**karyotes.
Puromycin	Binds to the A site, inhibits protein synthesis irreversibly.
Diphtheria toxin	Inhibits protein elongation by inactivating EF2
Actinomycin D	Inhibits DNA-dep-RNA synthesis
Rifamycin B (Rifampin)	Inhibits DNA-dep-RNA synthesis (Inhibits transcription) Acts on β-subunit of RNA polymerase.
Amanita phylloides	Inhibits RNA synthesis by blocking mRNA production. (inhibits RNA polymerase II)

OTHER MEDICATIONS:

Methotrexate
Analog of folic acid that is a competitive inhibitor of DHFR (dihydrofolate reductase). Therefore, it inhibits deoxythymidylate synthesis.

Aminopterin
Analog of folic acid, and competitive inhibitor of DHFR (dihydrofolate reductase). Therefore, it inhibits deoxythymidylate synthesis.

5-Fluorouracil
This drug inhibits *thymidylate synthase* (enzyme for dUMP $\longrightarrow$ dTMP). It inhibits deoxythymidylate synthesis.

Eukaryotic RNA Polymerase	RNA type	% of RNA	Synthesized In	Sensitivity to α-amanitin
pol I	rRNA	80%	nucleolus	(Not effective)
pol II	mRNA	8%	nucleoplasm (nucleus)	**VERY sensitive**
pol III	tRNA	12%	nucleoplasm (nucleus)	Fairly sensitive

Carbohydrates

CARBOHYDRATE STRUCTURE

MONOSACCHARIDES

General structure: $(CH_2O)_n$ where $n = 3, 4, 5, 6$, etc. . .

Name	# of carbons	Example	Formula
Triose	3	Glyceraldehyde	$C_3H_6O_3$
Tetrose	4	Threose, Erythrose	$C_4H_8O_4$
Pentose	5	Ribose, Xylose, Xylulose	$C_5H_{10}O_5$
Hexose	6	Glucose, Galactose, Fructose, Mannose	$C_6H_{12}O_6$

Aldoses (Glucose, Galactose, and Mannose) are monosaccharides which contain an *aldehyde*.

General structure:

$$(CH_2O)_n - \underset{\underset{H}{|}}{C} = O$$

Glucose is an Aldohexose (the aldehyde group is needed for reduction to occur):

	Carbon #	
Aldehyde → C—H	1	C—H
H—C—OH	2	H—C—OH
HO—C—H	3	HO—C—H
H—C—OH	4	H—C—OH
H—C—OH	←5→	HO—C—H
H—C—OH	6	H—C—OH
H		H
D-Glucose		L-Glucose

Ketoses (like Fructose) are monosaccharides which contain ketones.

General structure:

$$HO—\overset{\displaystyle H}{\underset{\displaystyle H}{C}}—\overset{\displaystyle O}{C}—(CH_2O)_n$$

Fructose is a ketohexose:

	Carbon #	
H—C—OH	1	H—C—OH
Ketone → C=O	2	C=O
HO—C—H	3	HO—C—H
H—C—OH	4	H—C—OH
H—C—OH	←5→	HO—C—H
H—C—OH	6	H—C—OH
D-Fructose		**L-Fructose**

L or D Configuration

The position of the hydroxyl on the second to last carbon of a monosaccharide (carbon #5 in hexoses) indicates whether that monosaccharide is in the L or D configuration. The last hydroxyl is on the *left* in **L**-carbohydrates and on the *right* in **D**-carbohydrates. *Enantiomers* are mirror images of each other. Carbohydrates (sugars) are usually in the **D** form. Amino acids are usually in the **L** form.

OPTICAL ACTIVITY

Carbohydrates can have many asymmetric chiral positions (i.e., carbons 2, 3, 4, and 5 in glucose; therefore, 4 chiral positions) so they are optically active. Compounds that *rotate light* to the left are designated "l" and to the right "d" (dexter is the Latin word for right). This should not be confused with L + D configurations.

Isomers and Epimers

Isomers: Have same chemical or molecular formula, i.e., $C_6H_{12}O_6$.
Epimers: Are isomers that differ only at one position:

$$
\begin{array}{c}
\text{CHO} \\
| \\
\text{—C—OH} \\
| \\
\text{HO—C—} \\
| \\
\text{—C—OH} \\
| \\
\text{—C—OH} \\
| \\
\text{CH}_2\text{OH}
\end{array}
\qquad \longleftrightarrow \qquad
\begin{array}{c}
\text{CHO} \\
| \\
\text{HO—C} \\
| \\
\text{HO—C—} \\
| \\
\text{—C—OH} \\
| \\
\text{—C—OH} \\
| \\
\text{CH}_2\text{OH}
\end{array}
$$

D-Glucose   D-Mannose

The above example of epimers (Mannose and Glucose) differ at C-2. Another example would be Galactose and Glucose—which differ at C-4. The enzyme, *epimerase*, allows the conversion between the two sugars.

Stereoisomers or Enantiomers:

Isomers that differ at more than one position. Enantiomers are mirror images of each other. Monosaccharides can have up to 2N stereoisomers (N = the number of chiral positions), unless they have a plane of symmetry.

Ring Forms (Anomers):

$$
\begin{array}{c}
\overset{6}{\text{CH}_2\text{OH}} \\
| \\
\text{CH—O} \\
|/_5 \quad \backslash| \\
_4\text{C OH} \quad \text{C }_1 \\
|\backslash| \quad _2 \;|/| \\
\text{HO C——C OH} \leftarrow \\
_3| \qquad | \\
\text{OH}
\end{array}
\qquad\qquad
\begin{array}{c}
\overset{6}{\text{CH}_2\text{OH}} \\
| \\
\text{CH—O OH} \leftarrow \\
|/_5 \quad \backslash| \\
_4\text{C OH} \quad \text{C }_1 \\
|\backslash| \quad _2 \;|/| \\
\text{HO C——C} \\
_3| \qquad | \\
\text{OH}
\end{array}
$$

α-D-(+)-Glucose   β-D-(+)-Glucose
(*downward* OH)   (*upward* OH)

(Small numbers indicate number assigned to each carbon)

Hydroxyls on the **right** in the linear structure are **down** on the rings.
Hydroxyls on the **left** in the linear structure are **up** on the rings.

The ring form is the predominant form that carbohydrates take in the real world. If you put a sample of α-glucose molecules into solution some of the molecules would spontaneously **mutarotate** and equilibrate with the β-form, and to a very small extent a non-ring linear form.

Anomers

When sugars form a ring, the aldehyde or ketone carbon becomes asymmetric and is called "**anomeric**." The **α and β forms** of a monosaccharide are called **anomers**. If a sugar's anomeric site can be oxidized to a carboxyl, it may reduce another molecule so it is called a **reducing sugar**.

POLYSACCHARIDES (POLYMERS)

Disaccharides :
Are 2 monosaccharides joined by a glycosidic bond.

Sucrose = fructose + glucose (sucrose is in table sugar)
Lactose = galactose + glucose (lactose is in milk)
Maltose = glucose + glucose (maltose is in beer)

Oligosaccharides:
Are several monosaccharides joined by glycosidic bonds.

Polysaccharides are many monosaccharides joined by glycosidic bonds.

2 types of **Glycosidic bonds**:

1. α-bond:

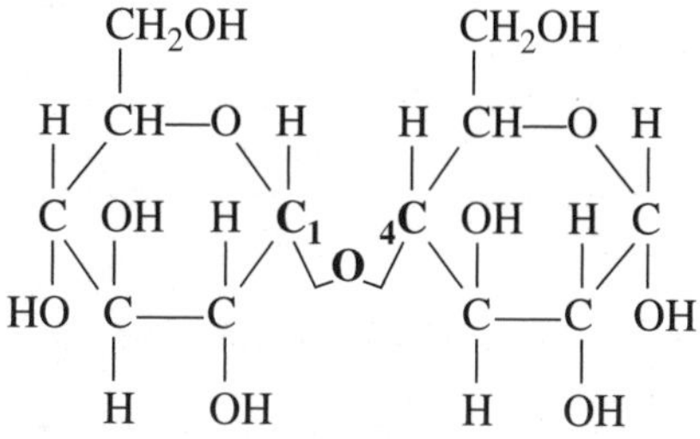

α-Bond

2. β-bond:

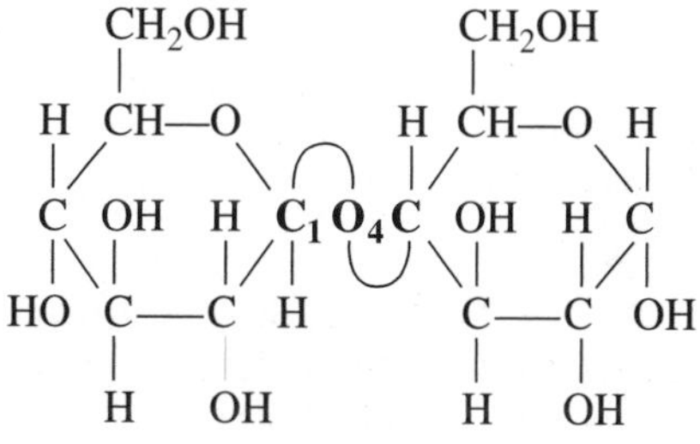

β-Bond

Carbohydrates have "O"- glycosidic bonds since 1 oxygen atom is joining the different monosaccharides (β-1,4-bond).

Starches are made up of glucose monosaccharides:

> **Glycogen** is the starch found in animals, it has many short branches; α-1,6 bonds occur at the beginning of each branch and α-1,4 bonds along the branches. Glycogen contains only α-anomers and α-bonds.

> **Cellulose**, the structural component of plants, is unbranched, linear, and has only β-1,4 bonds which cannot be broken by human enzymes (β-glucose polymer). Some animals have the special bacteria that have β-amylase.

> **Other Plant Starches** are mixtures of unbranched, linear, α-1,4 bond starches and starches similar to glycogen but with longer branches.

> Dietary starch is digested by salivary amylase, pancreatic amylase, maltase, and α-dextrinase (NOT by HCl, nor by lactase).

CLINICAL CORRELATIONS

SORBITOL

Sorbitol is *NOT* broken down in patients with diabetes (it is found in the lens, kidneys, and nerves). In diabetes, there is an increase in the glucose level. This increased Glucose is converted into an increased amount of Sorbitol by *aldose reductase*.

Sorbitol is converted by *sorbitol dehydrogenase* into Fructose.

When there is a defect in the enzymes (usually *aldose reductase*), this may cause *sterility* because fructose is needed for the sperm in seminal vesicles.

Glucose — *aldose reductase* ➤ Sorbitol — *sorbitol dehydrogenase* ➤ Fructose

(Therefore, if glucose increases, this increases sorbitol.)

In diabetes, check for *reducing sugars* (like glucose) in the urine. This is "spill over" from the blood and is filtered through the kidneys—overproduction and underutilization. Galactosemia and Fructosemia also cause *urine reducing sugars*.

The major dietary source of Fructose is through *honey*. Fructose is mainly used by the liver. It does NOT require insulin to be taken up into the cells (Glucose *does* require insulin). *Aldolase B deficiency* patients require fructose restriction from their diet—sucrase will not help the patient.

Glucose amount contributed by organs:
   LIVER > Kidney > skeletal muscle

CARBOHYDRATE DIGESTION

Carbohydrate digestion begins with *salivary* α-*amylase* in the mouth, and the enzyme can break α-1,4 bonds (found in starch). In the intestine, *pancreatic* α-*amylase* breaks α-1,4 bonds. Enzymes that are on the intestinal epithelial brush border include:

- α-*glucosidase* breaks α-1,4 bonds and also removes glucoses from the non-reducing ends.
- α-*Dextrinase* breaks α-1,6 bonds.
   Remember: Cellulose cannot be digested because of its β-1,4 bonds.
- Specific *disaccharidases*; i.e., maltase, sucrase, lactase etc., release the monosaccharides from disaccharides (Lactose → Galactose + Glucose).

Clinical correlation: Deficiencies of disaccharidase, such as lactase, produce intolerances that cause *diarrhea* because they draw water into the colon by osmosis. Furthermore, they produce *gas* (flatulence) when they are metabolized by bacteria in the colon to CO_2. Lactose intolerance is more common in Asians and Africans, and least common in northern Europeans. If you damage the intestine, you will have problems digesting disaccharides into monosaccharides. But, you will be able to degrade *starch* because of the salivary and pancreatic amylase. Lactose maldigestion may result from tissue destruction of the brush border, and can occur in children.

GALACTOSEMIA

Infant is unable to efficiently catabolize galactose (because of defective galactokinase), but can synthesize galactose from glucose. As a result, there is an increased galactose-1-phosphate. It is diagnosed by cord blood assay.

ABSORPTION

Monosaccharides are absorbed into intestinal epithelial cells by passive transport and then from the epithelial cells into the blood stream by Na^+/K^+ cotransport.

GLYCOLYSIS

Blood carries glucose to the cells of the body, where it is transported across the cell membrane by carrier molecules.

- This process is facilitated by insulin mainly in adipose and muscle tissues.
- Transport of glucose is **not** facilitated by insulin in red blood cells, intestinal mucosa, brain, and liver.
- Takes place in the cytoplasm.
- Produces *some* energy.

- Converts glucose into 3 carbon compounds which enter the Kreb's cycle.
- Glucokinase is specific for *glucose* and is a *liver* enzyme. While hexokinase is found in *every* cell.

There are three **ir**reversible reactions in Glycolysis:
1. Glucose — *hexokinase* ⟶ Glucose-6-Phosphate
2. Fructose-6-Phosphate — *phosphofructokinase* ⟶ Fructose-1,6-bisphosphate. (This is the regulating step or rate limiting reaction of glycolysis)
3. PEP — *pyruvate kinase* ⟶ Pyruvate

Glycolytic Pathways

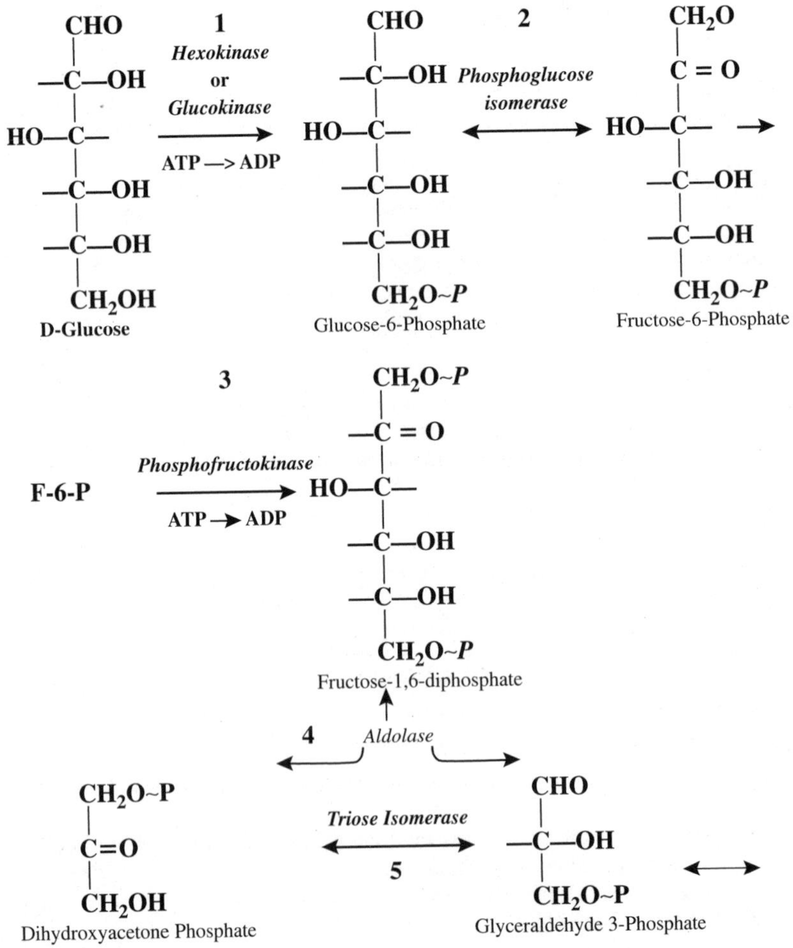

CHO | | —C—OH | | HO—C— | | —C—OH | | —C—OH | | CH₂OH
D-Glucose

1
Hexokinase
or
Glucokinase
$ATP \longrightarrow ADP$

CHO | | —C—OH | | HO—C— | | —C—OH | | —C—OH | | CH₂O~*P*
Glucose-6-Phosphate

2 *Phosphoglucose isomerase*

CH₂O | | C = O | | HO—C— | | —C—OH | | —C—OH | | CH₂O~*P*
Fructose-6-Phosphate

3

F-6-P
Phosphofructokinase
$ATP \longrightarrow ADP$

CH₂O~*P* | | —C = O | | HO—C— | | —C—OH | | —C—OH | | CH₂O~*P*
Fructose-1,6-diphosphate

4   *Aldolase*

CH₂O~P | | C=O | | CH₂OH
Dihydroxyacetone Phosphate

Triose Isomerase
5

CHO | | —C—OH | | CH₂O~*P*
Glyceraldehyde 3-Phosphate

6

Glyceraldehyde 3-Phosphate Dehydrogenase

⊢————————————————➤

$NAD^+ + P_i$ ———➤ $NADH + H^+$

$$\underset{\text{1,3-Diphosphoglycerate}}{\overset{\displaystyle \overset{O}{\|}}{C-O\sim P}}\quad\underset{}{\underset{}{H-C-OH}}\quad\underset{}{CH_2O\sim P}$$

7

Phosphoglycerate Kinase

————————➤

ADP ➤ ATP

$$\underset{\text{3-Phosphoglycerate}}{\overset{\displaystyle \overset{O}{\|}}{C-O^-}}\quad\underset{}{H-C-OH}\quad\underset{}{CH_2O\sim P}\;\longleftrightarrow$$

8

Phosphoglycero-mutase

⊢————————➤

$$\underset{\text{2-Phosphoglycerate}}{\overset{\displaystyle \overset{O}{\|}}{C-O^-}}\quad\underset{}{H-C-O\sim P}\;\longleftrightarrow\;\underset{}{CH_2OH}$$

9

Enolase

↘ H_2O

$$\underset{\text{Phosphoenol Pyruvate}}{\overset{\displaystyle \overset{O}{\|}}{C-O^-}}\quad\underset{}{\overset{\displaystyle \overset{O}{\|}}{C-O\sim P}}\quad\underset{}{CH_2}$$

10

Pyruvate kinase

————————➤

ADP ➤ ATP

$$\underset{\text{Pyruvate}}{\overset{\displaystyle \overset{O}{\|}}{C-O^-}}\quad\underset{}{C=O}\quad\underset{}{CH_3}$$

The above is *aerobic glycolysis*. In some cells with no mitochondria, red blood cells, and anoxic muscle tissue, **an**aerobic glycolysis may also occur. In **an**aerobic glycolysis, Pyruvate is then reduced to **Lactate**.

IMPORTANT STEPS IN GLYCOLYSIS

Regulatory Enzymes
These enzymes occur in irreversible reactions. They include: Hexokinase, Phosphofructokinase, and Pyruvate kinase, and regulate glycolysis.

Step # 1. Glucokinase and Hexokinase
- Glucokinase
  Found in the *liver*.
  Catalyzes *glucose* phosphorylation in the first step of glycolysis in the *liver*.
  Insulin facilitates its actions.
  High K_m and high V_m.
  Primary responsibility is to handle increased post-meal glucose surge.
  Catalyzed reaction is **ir**reversible.

- Hexokinase
  In *every cell* of the body.
  Less specific can phosphorylate different hexoses, i.e., mannose, fructose, glucose.
  Inhibited by glucose-6-phosphate.
  Has low K_m-high affinity so can continue phosphorylization even when low glucose levels are low.
  Catalyzes glucose phosphorylation, in first step of glycolysis.
  Catalyzed reaction is irreversible.
  Hexokinase is able to phosphorylate all: D-glucose, D-mannose, and D-fructose.

Step #2. Phosphofructokinase
Is the *rate limiting* reaction.
Actions are facilitated by high concentrations of AMP, and by fructose **2,6**-bisphosphate.
Allosteric inhibition by citrate, high concentrations of ATP, and by
Catalyzed reaction is irreversible.
Clinical correlation: You may misdiagnose a *fructokinase deficiency* (fructosuria) which is a "benign disease," for diabetes. This is because the reducing sugars will appear in the urine.

What you need to know about the other glycolytic enzymes not specifically discussed is that they are *reversible*, do **not** use or make energy, and are **not** regulatory. Just know the functions from the pathway diagram.

Epinephrine activates glycolysis and glycogenolysis in muscle. Epinephrine *inhibits* glycogen synthesis.

Important Reactions of Glycolysis		
Energy **Step #**	**Consumed or Produced**	**Reversibility**
1. *Hexokinase* **Glucokinase**	uses **1 ATP**	**ir**reversible
3. **Phosphofructokinase**	uses **1 ATP**	**ir**reversible

6. **Glyceraldehyde**
   3- Phosphate
   Dehydrogenase
	*makes **1 NADH + H**$^+$	reversible

7. **Phosphoglycerate** **Kinase**	makes **1 ATP**	reversible
10. **Pyruvate Kinase**	makes **1 ATP**	**ir**reversible
Net: +2 ATP (one Glucose becomes 2 Pyruvates)		
*NADH + H$^+$ later yields ATP by the electron transport chain.		

The glycolytic enzymes not in the chart are reversible and do not use or make energy.

GLUCONEOGENESIS

The production of glucose from organic precursors produced by glycolysis, or the degradation of amino acids, or glycerol from triglycerides. Many steps in gluconeogenesis make use of the reversibility of certain enzyme steps in glycolysis. This process occurs in the liver (90%) and the kidney (10%). The best way to learn gluconeogenesis is to memorize the steps that differ from the reverse glycolysis:

GLUCONEOGENESIS:

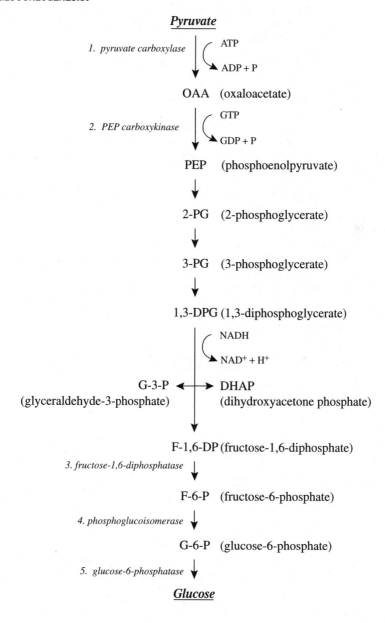

1. In the mitochondria, pyruvate is converted to oxaloacetate by *pyruvate carboxylase* (acetyl CoA increases the activity of this enzyme).

[Oxaloacetate is oxidized to malate and shuttled across the mitochondrial membrane into the cytoplasm where it is changed back to oxaloacetate.]
2. Oxaloacetate is converted to phosphoenolpyruvate (PEP) by *phosphoenolpyruvate carboxykinase*.

From this point glycolysis is reversed until fructose-1,6-bisphosphate is formed.

3. Fructose-1,6-bisphosphate is converted to fructose-6-phosphate by *fructose-1,6-bisphosphatase*.
4. Fructose-6-phosphate (using the reversible glycolytic enzyme, phosphoglucoisomerase) is converted to glucose-6-phosphate. Then, *glucose-6-phosphatase* cleaves the phosphate group, and produces glucose. *Glucose-6-phosphatase* is found in *liver* (not in skeletal muscle).

The synthesis of each molecule of glucose *consumes*: 4 ATP, 2 GTP, and 2 NADH.

GLYCOGEN
Contains branched chains of α-1,4-glycosidic links and α-1,6-glycosidic links. Made of **D**-glucose residues. Glycogen supplies most of the glucose released by the liver in the post-absorptive state. α-**1,4**-links produce the linear portion of glycogen, and the α-**1,6**-links produce the *branched* portion. Clinical correlation: normally, the ratio of α-1,4 to α-1,6 is 8 : 1; in a patient with a glycogen storage disease the ratio will increase to 1,000 : 1 (majority comprised of *linear* chain, rather than the branched chains).

GLYCOGENESIS

The initial step in *glycogen* synthesis is:
Glucose-**6**-phosphate—*phosphoglucomutase*→ Glucose-**1**-phosphate

Then,
Glucose-1-phosphate—*UDP-glucose pyrophosphorylase* → UDP-Glucose
("activated glucose")

UDP-Glucose—***Glycogen synthase***—(this enzyme adds glucose onto α-1,4 bonds, non-reducing ends)—***Glucosyl-(4:6)-transferase***—(this enzyme breaks the α-1,4 bond, and creates branches by making α-1,**6** bond) → GLYCOGEN.

GLYCOGENOLYSIS

The degradation of glycogen occurs in the cytosol by adding enzymes and water to remove the phosphates and glucose. The enzymes are considered "acid hydrolases" since they work in low pH.

The sequence of events is as follows:

Glycogen—*glycogen phosphorylase a*—(this enzyme adds phosphate and breaks off a glucose from the glycogen chain, leaving a minimum of four glucoses on the glycogen chain) → Glucose-1-phosphate + Glycogen

This glycogen is considered a "limit dextran." As a clinical correlation, *McArdle's disease* is a deficiency in *muscle* glycogen phosphorylase; therefore, there will be an accumulation of glycogen.

Glycogen then goes through a few more steps; the main enzymes include:
1. *Glucosyl (4:4) transferase*—(this debranching enzyme cuts all the glucoses off, but leaves 1 glucose to remain on the glycogen branch.)
2. *Amylo (1:6) glucosidase*—(this debranching enzyme adds water and the sequence continues to remove Glucose.)

Glycogen ➤ ➤ Glucose-1-phosphate—*hexokinase* ➤ Glucose-6-phosphate.

Note: Glycogenolysis (glycogen breakdown) will occur if there is an **increased** cAMP, Epinephrine, Ca^{2+}, active protein kinase, active phosphorylase kinase, active phosphorylase a, or a *Decreased* glucose level. The *first* product of glycogenolysis is Glucose-1-phosphate.

KETOSIS

Increased levels of ketones. Ketoacidosis can occur in starvation and uncontrolled diabetes mellitus (which mimics starvation because insulin deficiency does not allow glucose to enter many cells). These disease states result in deficiency of Kreb's Cycle intermediates and lower the blood pH. If untreated, ketoacidosis may result in death. Examples of ketone bodies include: Acetone, acetoacetate, and β-hydroxybutyrate. Ketone bodies are used as fuel by the *heart* (Ketone bodies are *NOT* used as fuel by the liver).

Ketones are *formed* in the liver:
2 Acetyl CoA ➤ Acetoacetyl CoA ➤ 3-Hydroxy-3-methylglutaryl CoA ➤ Acetoacetate ➤ β-hydroxybutyrate.

Oxidation is the loss of electrons (H^+) and **Reduction** is the *gain* of electrons.

ELECTRON TRANSPORT CHAIN (RESPIRATORY CHAIN)

Protons are transferred across the *inner* mitochondrial membrane, electrons flow through the chain starting with NADH, and end with the final electron carrier, oxygen. In mitochondria, the sites where **ATP is produced** are between:
1. NADH and flavoprotein,
2. Cytochrome b and cytochrome c,
3. Cytochrome c and cytochromes a, a3.

$$\text{NADH} \xrightarrow{\text{I}} \text{FMN} \xrightarrow{\text{II}} \text{CoQ} \to \textbf{Cytochrome chain}$$

Electrons pass through the *cytochrome* oxidase *chain*:

$$\text{(cytochrome) b} \xrightarrow{\text{III}} \text{c1} \to \text{c} \to \text{a} \xrightarrow{\text{IV}} \text{a3} \to \text{Oxygen}$$

CYTOCHROME

Hemoprotein that transports electrons or hydrogen. Four types: cytochrome a, b, c, and d. Cytochrome **a** has a large standard reduction potential. The *brain* has the highest use of cytochrome and therefore, cyanide can kill. (*See table below for other inhibitors of the cytochrome oxidase chain.*)

Enzyme	Complex	Electron acceptor	Inhibitors
NADH dehydrogenase	I	FMN (Fe-S)	Rotenone, Amytal
Succinate dehydrogenase	II	CoQ	Malonate
Cytochrome **b-c**$_1$ reductase	III	Cyto c$_1$	Antimycin A
Cytochrome oxidase	IV	1/2 O$_2$	Carbon monoxide, Cyanide, Sodium azide

ACTIVE TRANSPORT OF GLUCOSE INTO CELLS

Requires Na$^+$/K$^+$ ATPase active transport of Na$^+$. The method of transport is through sodium and glucose *symport*.

MATURE RED BLOOD CELL

Mature RBC's do not synthesize nucleic acids and proteins. They use reduced glutathione and glutathione peroxidase to metabolize H$_2$O$_2$. Glucose is metabolized by the glycolytic pathway (**an**aerobic glycolysis) *and* the pentose phosphate pathway (**PPP**). (The rbc does NOT form ATP by the electron transport system.) The PPP generates NADPH, and NADPH is used to maintain reduced glutathione—G6PD is necessary. PPP metabolizes only about 10% of the glucose. Lactate is an end-product of glucose metabolism.

GLUTATHIONE

Glutathione is a tripeptide, contains a sulfhydryl group (from Cysteine). Glutathione-SH is oxidized to G-S-S-G. NADPH-dependent glutathione reductase keeps it in a reduced state. In the rbc, glutathione (when in the reduced state) protects other sulfhydryl groups from oxidation. **G6PD deficient patients** may be affected by oxidizing agents like antimalarial *drugs*, aspirin, *infection*, and *fava beans* that will lead to hemolysis of red blood cells and eventually *hemolytic anemia*. Old rbc's are very vulnerable, and young cells are left. Therefore, on the second exposure, there is *less* of an effect. Dietary anti-oxidants include Vitamin C, Vitamin E, and β-carotene (Vitamin A); Hydrogen peroxide is detoxified by the following enzymes: 1. Catalase, 2. Superoxide dismutase, and 3. Glutathione peroxidase.

PENTOSE PHOSPHATE PATHWAY
(PPP or Hexose monophosphate shunt)

The pentose phosphate pathway acts to supply **NADPH** and **pentoses (5-carbon sugars)** for nucleic acid synthesis. It also converts pentoses to hexoses and trioses. NADPH is the reducing equivalent for biosynthetic reactions. The PPP is the only producer of NADPH in red blood cells.

It also forms most reducing equivalents (NADPH) used in Fatty Acid synthesis. In the irreversible first step, G-6-PDH (Glucose-6-phosphate dehydrogenase) is the regulatory enzyme that is positively regulated by $NADP^+$. The enzymes of the PPP are in the cytosol. In the PPP schematic below, we begin with a **6**-carbon glucose, and lose the carbon #1 (as CO_2) to eventually create ribose-**5**-phosphate.

PENTOSE PHOSPHATE PATHWAY:

$$NADP \quad NADPH + H^+$$

Glucose-6-Phosphate ⟶ 6-phosphoglucono-δ-lactone

6-Phosphogluconate

$$NADP$$

$$NADPH + H^+ \qquad CO_2$$

Ribulose-**5**-Phosphate

Ribose-5-P

Ribose-5-P + Xylulose-5-P ◄─► Glyceraldehyde-3-P + Sedoheptulose-7-P

Sedoheptulose-7-P + Glyceraldehyde-3-P ◄─► Erythrose-4-P + Fructose-6-P

Xylulose-5-P + Erythrose-4-P ◄─► Glyceraldehyde-3-P + Fructose-6-P

NET:

2 Xylulose-5-P + Ribose-5-P ◄─► Glyceraldehyde-3-P + 2 Fructose-6-P

Enzymes include:
Isomerase, transketolase (requires thiamine pyrophosphate), transaldolase, epimerase.

PPP is a very active pathway in tissues that synthesize lipids or steroids (adipocytes, adrenal cortex, lactating mammary glands, and the liver). Percentage of metabolized glucose by this pathway varies for different tissues of the body. Reactions occur in the *cytoplasm* of the cell. CO_2 is generated from glucose in the red blood cell from this pathway. PPP activity is very *low* in skeletal muscle.

CITRIC ACID CYCLE (TCA, KREB'S CYCLE)

NOTE: Memorize *everything* in this section. This is a short section and it will likely be covered on your exam.

Citric Acid Cycle (TCA, Kreb's Cycle)

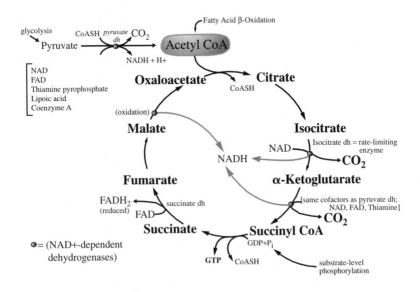

The Citric Acid Cycle is *inhibited* by an **an**aerobic environment, arsenite (or other inhibitors), malonate, and fluoroacetate (NOT fluorouracil).

FAD acts as a cofactor for the oxidation of succinate. FAD (*flavin* adenine dinucleotide) is the prosthetic group of succinate dehydrogenase.

Converting succinate directly to oxaloacetate produces a maximum of **5** ATP's. (Not going through citrate—2 ATP's come from $FADH_2$, and 3 ATP's come from NADH.) The total number of ATP produced comes from:

$$3 \text{ NADH} \times 3 = 9 \text{ ATP} \text{———(Malate-Aspartate Shuttle)}$$
$$+ 1 \text{ FADH}_2 \times 2 = 4 \text{ ATP} \text{———(Glycerol-3-phosphate/DHAP Shuttle)}$$
$$\underline{+ 1 \text{ GTP} \approx 1 \text{ ATP}}$$

This is 12 ATP equivalents per Acetyl CoA
$\times$ 2 Acetyl CoA/glucose = **24 total ATP**.

There are **three** NAD+-dependent dehydrogenases: Isocitrate dehydrogenase, α-Ketoglutarate dehydrogenase, Malate dehydrogenase.

Pyruvate dehydrogenase and *α-Ketoglutarate dehydrogenase* require the following cofactors in the mitochondrial complex: CoA (Coenzyme A), NAD+, FAD, Lipoic acid, and thiamine pyrophosphate.

The vitamin B complex is important in the *oxidation of malate*, and for the *decaroxylation* of *α-ketoglutarate*. Furthermore, it is important in creating *Acyl-CoA intermediates*.

Remember, the *Malate-Aspartate Shuttle* occurs in the liver, kidney, and heart to generate **3 ATP/NADH**. The *Glycerol-3-phosphate/DHAP Shuttle* occurs in skeletal muscle and the brain to generate **2 ATP/FADH$_2$**.

THE CORI CYCLE

Exercising muscles converts glucose from the blood into lactate. This lactate then enters the blood, and is changed by the liver back to glucose (and the cycle repeats itself).

The Cori Cycle

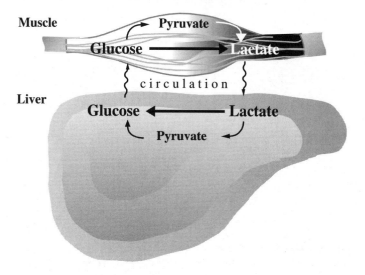

GENERAL STRUCTURE OF LIPIDS

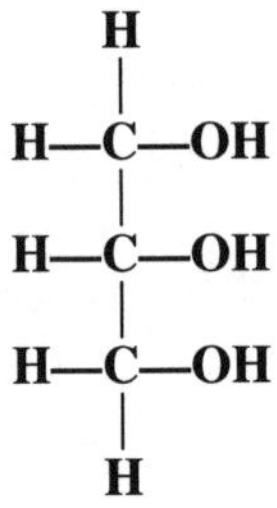

$$
\begin{array}{c}
H \\
| \\
H-C-OH \\
| \\
H-C-OH \\
| \\
H-C-OH \\
| \\
H
\end{array}
$$

Glycerol

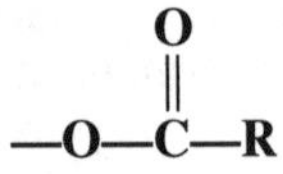

$$-O-\overset{\overset{\displaystyle O}{\|}}{C}-R$$

Fatty Acid (FA)

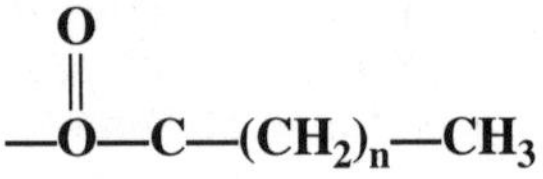

$$-O-\overset{\overset{\displaystyle O}{\|}}{C}-(CH_2)_n-CH_3$$

Saturated Fatty Acid

$$-O-\overset{\overset{\displaystyle O}{\|}}{C}-[(CH_2)_n-CH=CH-(CH_2)_n]_n-CH_3$$

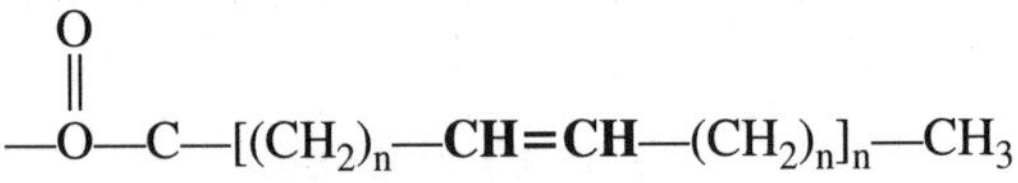

Unsaturated Fatty Acid

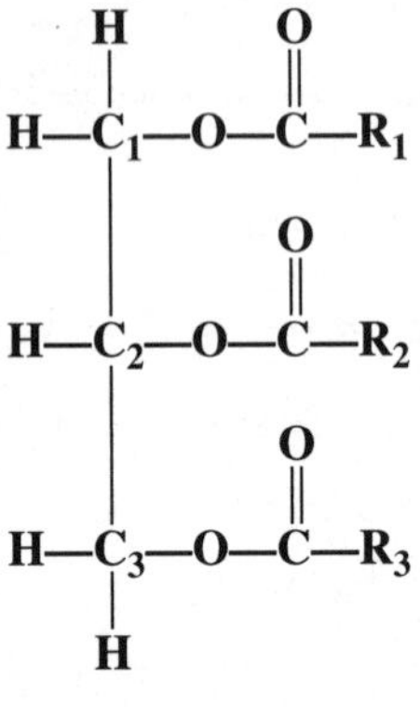

$$
\begin{array}{c}
H \qquad O \\
| \qquad \| \\
H-C_1-O-C-R_1 \\
| \qquad\quad O \\
| \qquad\quad \| \\
H-C_2-O-C-R_2 \\
| \qquad\quad O \\
| \qquad\quad \| \\
H-C_3-O-C-R_3 \\
| \\
H
\end{array}
$$

Triacylglycerol (TG)

CIS- VERSUS TRANS-FATTY ACIDS

The more *trans*-fatty acids, then the worse off you are for health (thick, hard fat)! Remember, the "softer" the butter—the better the fat. This is why olive oil is a "good" oil, since it has the *least* amount of *trans*-fatty acids.

SATURATED FATTY ACIDS

Saturated fatty acids occur naturally as *cis*-fatty acids. They have **no** double bonds and almost all the carbons have 2 hydrogens attached to them, so they are "saturated with hydrogens". Saturated fatty acids include: Stearic acid, and

Lauric acid. Furthermore, *Palmitic acid* (16 carbons) and *stearic acid* (18 carbons or 18:0) are the major human saturated fatty acids. (When we write 18:0 this means 18 = # of Carbons, and 0 = the number of double bonds.)

MONO-UNSATURATED FATTY ACIDS

These fatty acids have **1** carbon-carbon double bond (**—CH=CH—**). *Palmitoleic acid* (16:1, with the double bond at carbon #9), and *Oleic acid* (18:1) are the major human mono-unsaturated FA's. Many plants and vegetable oils are mono-unsaturated fatty acids, and therefore are good for you.

POLY-UNSATURATED FATTY ACIDS

These fatty acids have *more than* one carbon-carbon double bond. Examples include: *Linoleic acid* (18:2), Linolenic acid (18:3), and arachidonic acid (20:4). Linoleic acid and Linolenic acid are considered *essential fatty acids*, and must be taken in from the diet. Arachidonic acid is a precursor for prostaglandins and leukotrienes. Arachidonic acid may be essential if there is limited dietary linoleic acid—since linoleic acid forms arachidonic acid.

ARACHIDONIC ACID

Arachidonic acid is synthesized from *essential* fatty acids. It forms eicosanoids (20 carbons) like: Prostaglandins, leukotrienes, and thromboxanes. Essential fatty acids (like linoleic acid from the diet) forms linolenic acid which forms Arachidonic acid. Arachidonic acid is released from the membrane by phospholipase **A2**. (*See diagram below*)

Prostaglandin & Leukotriene Synthesis

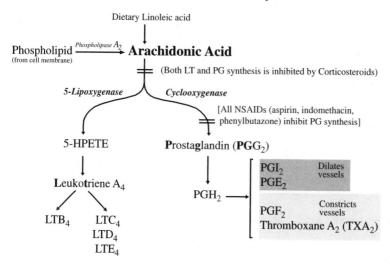

Aspirin inhibits the enzyme *Cyclooxygenase*. It therefore inhibits the synthesis of prostaglandins and thromboxanes. As you can see from the diagram, it does NOT inhibit *lipoxygenase* nor the synthesis of leukotrienes.

DIGESTION

Dietary triglycerides enter the small intestine where bile salts—produced by the liver and stored in the gallbladder—are secreted and emulsify the triglycerides. Then, *pancreatic lipases* cleave the triglycerides into free fatty acids and 2-monoglycerides. These hydrophobic compounds aggregate into small droplets called "micelles" which are absorbed at the intestinal microvilli. Bile salts are then reabsorbed in the ileum and then reprocessed in the liver.

2-Monoglyceride

$$
\begin{array}{c}
H \\
| \\
H\!-\!C_1\!-\!OH \quad O \\
| \qquad\qquad \| \\
H\!-\!C_2\!-\!-O\!-\!C\!-\!R \\
| \\
H\!-\!C_3\!-\!OH \\
| \\
H
\end{array}
$$

LIPIDS

In fat tissue, *triacylglycerol* (TG) is *synthesized in fat cells* from precursors. It is a fatty acid ester of glycerol. TG in the chylomicron is degraded by lipoprotein lipase (from adipose tissue) into free fatty acids (FFA's) and glycerol. The FFA of TG is the major energy store of the body. Glycerol is used by the liver and forms glycerol-3-phosphate.

The bulk of *dietary* lipids are triacylglycerol. Lipids are part of the membrane structure, and are **in**soluble in water or *hydrophobic*. They are transported with protein in the body as "lipoprotein particles" (i.e., chylomicrons, VLDL, LDL, and HDL). In lipid digestion, lipids in the duodenum are emulsified by "detergents" or *bile salts*. The main function of bile acids is the formation of *micelles*. Bile salts help in the absorption of cholesterol. Fatty acids with a chain length *less than 12* carbons will pass through the mucosal cell *directly* into the blood stream. The primary site for lipid absorption is at the *mucosal cells of the intestine*.

Lipases break down lipids (TG's) to release the fatty acids. *Lingual lipase* degrades TG's in the mouth. *Gastric lipase* degrades short to medium-chain lipids, and works at a pH of 4 to 5—which is the pH of a *baby's stomach*, but not very useful in adults. When a food bolus reaches the small intestine, *pancreatic lipase* acts as an esterase and removes fatty acids (from C #1 and C #3) of the triacylglycerol. Furthermore, *phospholipase A_2* removes a fatty acid off C#2, and *cholesterol esterase* breaks fatty acids off of cholesterol. *Colipase* is secreted in the pancreatic juice. Colipase is a small protein that stabilizes pancreatic lipase.

An increase in cAMP activates *triglyceride lipase* and regulates lipolysis. *Hormone sensitive lipase* is an enzyme that mobilizes triacylglycerols from adipose tissue.

Mixed micelles contain bile salts, 2-monoacylglycerol, and FFA's.

Chylomicrons are much larger than mixed micelles and are made of approximately 90% TG, Vitamins A, D, E, and K, phospholipids, cholesterol esters, and apoprotein B-48. Chylomicrons have a hydrophilic charged surface and are released by exocytosis into the lymphatics (chylomicrons that are formed in the intestinal mucosa). Chylomicrons are made in the intestinal epithelial cells when an individual eats a meal. They have the *highest* triacylglycerol content (>80%TG)—they transport dietary triglycerides. They are a substrate for lipoprotein lipase. *(See illustration below for schematic chylomicron formation.)*

Clinical correlation: *Malabsorption* of lipids results in *steatorrhea*, because the excess lipids remain in the feces. This gives the stool a greasy appearance and the stool will float. The cause of this can include liver disease (not enough bile is being produced), gall stones (the bile is occluded), pancreatic disease (enzyme disorder), or intestinal mucosal cell defect.

Also, not all "fat" is bad. Notice: "fat" is needed for proper absorption of lipid-soluble vitamins as well as other functions.

Chylomicron Formation

Apoprotein B-48

2– Monoacylglycerol + 2 fatty acyl CoA ＊→ Triacylglycerol

Chylomicron

Cholesterol + Fatty acyl CoA ＊→ Cholesterol Ester
(cholesteryl ester)

Fatty acids

Phospholipids
(A,D,E,K)

intestinal mucosal cell

To Lymphatics

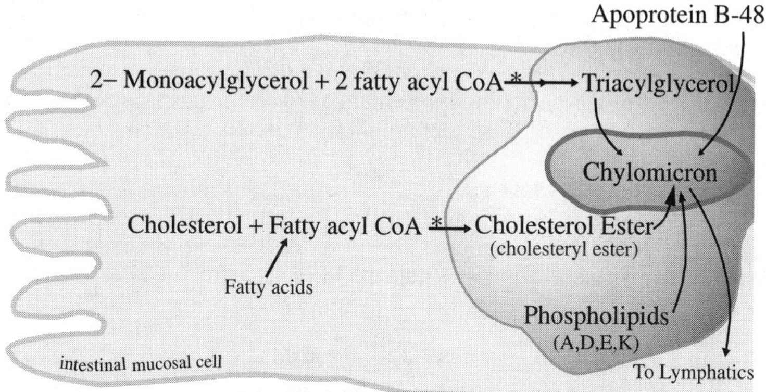

*Acyl Transferase

GREATEST TO LOWEST DENSITY
LDL — IDL — VLDL (TG) → Chylomicron
The lowest density means the highest amount of triglycerides and the largest size (increased fat, decreased protein).

LDL
Lipoprotein lipase action on VLDL or IDL will form LDL. LDL *transport* cholesterol **to** *peripheral tissues*. A cause of hypercholesterolemia is an LDL receptor deficiency. LDL is synthesized in the *serum and plasma*, NOT the liver. LDL contains mainly *cholesterol* and *cholesterol esters*.

HDL
HDL is important in *removing* cholesterol **from** *peripheral tissues*. HDL is made in the *liver*. HDL contains an apolipoprotein that activates lecithin cholesterol acyl transferase (HDL is associated with LCAT). *Tangier's disease* has a deficiency of this lipoprotein.

LECITHIN CHOLESTEROL ACYL TRANSFERASE (LCAT)
LCAT converts cholesterol to a cholesterol ester. LCAT is activated by apoprotein-A I, an important part of HDL (not VLDL/ LDL).

APOPROTEIN B 48
This is needed to package the TG for forming chylomicrons in the small intestine.

APOLIPOPROTEIN B-100
This is made in the liver, and is a portion of VLDL. B-100 receptor deficiency would increase blood cholesterol.

APOPROTEIN C II
This activates extracellular *lipoprotein lipase*, and degrades TG's into FFA's and Glycerol.

LIPOPROTEIN LIPASE
Lipoprotein lipase acts on chylomicrons. It requires (and is activated by) apolipoprotein C. The enzyme is found attached to capillaries. Exogenous hypertriglyceridemia is caused by lack of *lipoprotein lipase*; causing a severe chylomicronemia. Familial dysbetalipoproteinemia (hypertriglyceridemia) is due to deficient *apoprotein E*; causing increased chylomicron remnants.

FAMILIAL HYPERCHOLESTEROLEMIA
This is an autosomal dominant disorder (deficient LDL receptors). The *decreased* LDL plasma membrane *receptors* results in *increased* LDL and *cholesterol* levels; therefore, these patients have an increased risk of coronary artery disease.

Cholesterol —— (*desmolase*) ⟶ pregnenolone ⟶ Progesterone ⟶
forming aldosterone, cortisol, and sex hormones.

CHOLESTEROL SYNTHESIS
The rate-controlling step (rate-limiting step) is:
3-**H**ydroxy-3-**m**ethylglutaryl **CoA** ⟶ Mevalonic Acid
HMG-CoA reductase

Cholesterol synthesis requires molecular oxygen.

HMG-CoA REDUCTASE
This enzyme is involved in the rate-limiting step of cholesterol synthesis. It is *inhibited* by cholesterol or Lovastatin, and *activated* by insulin.

SQUALENE
Mevalonic acid is phosphorylated into Squalene, which forms Lanosterol and finally, Cholesterol.

FATTY ACID SYNTHESIS

Short-term control is by acetyl CoA carboxylase and fatty acid synthase. Long-term regulation is by diet and induction of enzymes. Fatty acid synthesis occurs in the cytosol, and oxidizes fatty acids in the mitochondria for energy. Citrate "carries" the fatty acid out of the mitochondria. (*See diagram below.*)

Fatty Acid Synthesis
(making Palmitate C16)

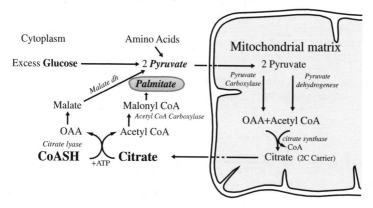

DE NOVO: FATTY ACID SYNTHESIS
Acetyl-CoA carboxylase is the rate-limiting step. Synthesis occurs in the cytoplasm. Acyl-carrier protein and NADPH are required.

1. **Synthesis of cytoplasmic acetyl CoA.** Transfer of acetate from acetyl CoA *from the mitochondria to the cytoplasm*. (CoA cannot cross the mitochondrial membrane). This is called *translocation of citrate*. It occurs if isocitrate dehydrogenase is inhibited (by an increase in ATP and citrate).

2. **Acetyl CoA carboxylated to malonyl CoA.** *Acetyl CoA carboxylase* (with *biotin* as a cofactor); this is the rate limiting step and is stimulated by *citrate*. Malonyl CoA or the end product palmitoyl CoA will decrease the activity of acetyl CoA carboxylase.

3. **Fatty acid synthase** acts as a complex. Acetyl CoA-ACP transacylase transfers acetate off the acetyl CoA to the acyl carrier protein (ACP). Joins *cysteine*, then *malonate* from malonyl CoA binds to the ACP, goes through reductase, dehydratase, and another reductase reaction and forms a 4-carbon chain. The repetition adds 2-carbons from the

malonyl CoA, and the fatty acid chain grows (i.e., 16 carbon chain = palmitate; saturated FA).

Citrate is the carrier of acetyl groups *out* of the mitochondria and allows the continuation of lipogenesis.

Fatty acid synthesis uses reduced **NADPH** in fatty acid synthesis derived from the malate-citrate shuttle (malate to pyruvate), and from the pentose phosphate pathway. NADPH is also required for steroid biosynthesis in the adrenal gland. It acts as a cofactor of fatty acid *synthesis*.

Coenzyme A (CoA) links to intermediates in fatty acid oxidation. Remember, the inner mitochondrial membrane is impermeable to CoA.

FA's are transported bound to *serum albumin*. They are transported from adipose tissue to: the muscle, liver and kidney.

FA's are stored as mono-, di-, and tri-acylglycerols (triglycerides); which are FA molecules esterified with glycerol.

Malonyl-acyl carrier protein is the precursor form used in the elongation reaction producing palmitate.

Overall FA Synthesis:
Condensation ➔ *Reduction* ➔ Dehydration ➔ *Reduction.*

FATTY ACID DEGRADATION

Oxidation is the first reaction. Overall FA Degradation:
Oxidation ➔ Hydration ➔ *Oxidation* ➔ Thiolysis.

The major fuel storage = FA's in fat tissue. Acetyl CoA is the final product of the path.

Hormone-sensitive lipase (h.s.l.) is activated by *cAMP-dep protein kinase*, and removes FA from C#1 or C#3 of the TG:
Adenylate cyclase is **activated** by glucagon and epinephrine. Adenylate cyclase is **inhibited** by insulin and glucose. Activation causes an increase in cAMP which activates protein kinase. Protein kinase activates h.s.l. in the presence of ATP; (phosphatase inactivates h.s.l.).
H.s.l. can then form fatty acid and diacylglycerol (from a triacylglycerol).

β-OXIDATION
This is FA degradation or the catabolic pathway of FA's which removes 2 C fragments from the carboxyl end and forms acetyl CoA (by *oxidizing the β-carbon*). Each cycle of fatty acid oxidation *yields*: **Acetyl CoA**, Fatty acyl-CoA, **FADH$_2$**, and **NADH**.

β-Oxidation of fatty acids *requires* these substrates (cofactors): FAD, NAD, and CoA.

β-Oxidation occurs in the *mitochondria*, and the FA needs to be transported across the inner mitochondrial membrane, but the bulky CoA cannot get across and must be "carried" across by carnitine.

Carnitine acyl transferase I (CAT)
This is the rate-limiting step in fatty acid oxidation.

Carnitine is the *carrier* that transports acyl groups in the "carnitine shuttle". It is involved in trans*acyl*ation. It is an amino acid involved in lipid metabolism, and required for the transport of **long** chain fatty acids *from* the cytosol *into* the mitochondria. Carnitine deficiency results in decreased fatty acid oxidation.

Malonyl CoA
Inhibits *carnitine acyltransferase I*. (Thus, it cannot transfer new FA into the mitochondria.) It is formed in the *cytoplasm* by acetyl CoA carboxylase. When malonyl CoA is increased, it causes a *decrease* in β-oxidation of *long* chain fatty acids (greater than 12 carbons).

OXIDATION OF FA'S OF *EVEN*-NUMBER OF CARBONS (N):
   $(n/2) - 1$
Calculate the ATP yield using this short-cut:
1. from FA-CoA to Acetyl CoA (3 ATP/NADH) $[(n/2) - 1] \times 3$
2. from FA-CoA to Acetyl CoA (2 ATP/FADH$_2$) $[(n/2) - 1] \times 2$
3. from Acetyl CoA to CO$_2$ and H$_2$O (TCA cycle) $(n/2) \times 12$
4. loss from breaking energy bonds (thiokinase reaction) $\underline{-2}$
                                                    Total ATP yield

i.e., palmitoyl-CoA (which is a 16 Carbon FA) would yield:
   1. $(16/2 -1) \times 3 = 21$
   2. $(7) \times 2 = 14$
   3. $(8) \times 12 = 96$
   4. $\underline{-2}$
      total = 129 ATP yield

OXIDATION OF FA'S OF *ODD*-NUMBER OF CARBON CHAINS:
Same, but the last step forms a 3-C compound, *propionyl CoA*.
Calculate this way:

1. Through the ETC (2 from FADH2, 3 from NADH) $\left[\dfrac{(n-1)}{2} - 1\right] \times 5$

2. Going through the TCA cycle $\dfrac{n-1}{2} \times 12$

3. must subtract 7 since coming into TCA cycle late -7
   (only make 5 instead of 12 ATP's)
4. loss from origination (thiokinase) $\underline{-2}$
                                        Total ATP yield

i.e., an 11 Carbon FA will yield:

$$\frac{11-1}{2} \times 12 = 60$$
$$(5-1) \times 5 = 20$$
$$-7$$
$$\underline{-2}$$
71 total ATP yield from an 11 carbon FA.

α-OXIDATION
Would build-up *phytanic acid* if lacking. α-Oxidation of fatty acids occurs at the *smooth endoplasmic reticulum.*

OTHER LIPID TOPICS

BILE ACIDS
Cholic and chenodeoxycholic acids are *primary* bile acids. Deoxycholic and lithocholic acids are *secondary* bile acids. Bile acids are important in digestion. They are synthesized in the liver from cholesterol. Bile acids + Glycine or Taurine forms Bile *Salts*. Bile salts are secreted in bile to the duodenum and go back to the liver. Bile contains bile salts and bile emulsifies lipids. Bile is the number one way to remove cholesterol from the body. Lung surfactant is *dipalmitoyl phosphotidyl choline*, which decreases the surface tension in the lungs. In premature infants, this is not formed and the infant has difficulty breathing because of the lack of type **II** pneumocyte production.

TAUROCHOLATE
Taurocholate is involved in the digestion of dietary fat in the small intestine. In the *liver*, cholesterol forms the *primary* bile acids: cholic and chenodeoxycholic acids. In the *intestine*, lithocholic and deoxycholic acids (secondary bile acids) are formed from the primary acids. Then, in the *liver* the bile acids are conjugated with glycine (to form glycocholate—a bile salt), and taurine (to form taurocholate—a bile salt).

CEREBROSIDE
This is a simple glycolipid (single sugar bound to lipid). It is made of a sphingosine and an FA chain, plus one glucose or one galactose.

SPHINGOSINE
This is the backbone of ganglioside, cerebroside, ceramide and sphingomyelin.

SPHINGOMYELIN
A sphingolipid, sphingomyelin comes from *sphingosine + FA + ceramide.*

SPHINGOLIPID SYNTHESIS
Pyridoxal phosphate and NADPH are *required* co-factors. Palmitoyl-CoA and Serine are precursors. In formation of complex sphingolipids, ceramide is a major intermediate. Sphingolipids are mainly found in the brain and in nerve

tissue. Tay-Sach's Disease, a lipid-storage disease, is a defect in the *lysosomal* enzyme *hexosaminidase A*.

CERAMIDE

A ceramide is a sphingosine plus an FA (i.e., an 18 carbon sphingosine attached to a fatty acid by an amide bond). Ceramide is an integral part in the formation of sphingolipids. It is an intermediate in the formation of gangliosides, sulfatides, and sphingomyelin (NOT with Diacylglycerol).

PHOSPHOLIPIDS

Phospholipids are amphipathic, and all have phosphate. They serve an important membrane function and in bile and surfactant. May be synthesized de novo from CDP-diacylglycerol and a polar alcohol. Have a polar head made of an amino alcohol or a sugar alcohol. (It is NOT an important storage form of lipid.) Phospholipase A_2 degrades phospholipids and is inhibited by glucocorticoids. Phospholipase C is an enzyme that degrades the membrane and releases the second messenger, and it attacks the phosphatidyl inositol.

PHOSPHOLIPID EXCHANGE PROTEINS

Move phospholipids between the *intracellular membranes*.

GLYCOLIPIDS

These are membrane components like a ganglioside. It is made from an FA + sphingosine to yield a ceramide. The ceramide joins with a sugar to form a cerebroside. The cerebroside joins with glucose to become a "glucose cerebroside."

BROWN FAT

Brown fat may *produce heat*. The excess calories of the diet may induce heat production.

Vitamins and Hormones

VITAMINS

Vitamins are received from the diet and are needed for growth.

FAT-SOLUBLE VITAMINS

Vitamins A, D, E, and K. Fat-soluble vitamins are stored and therefore, difficult to metabolize.

VITAMIN A

Three forms:
Retinol (vitamin A alcohol and the transport form),
Retinal (vitamin A aldehyde),
Retinoic acid (vitamin A acid; excites growth and epithelial cells).

In vision, retinal combines with opsin and forms the pigment, rhodopsin.

Cis-retinal changes to trans-retinal on exposure to light, with release of opsin.

In the retina, trans-retinal is reduced to trans-retinol.

Then, in the liver it changes to *cis*-retinol.

Cis-retinol goes back to the retina and becomes Cis-retinal again, which then binds with opsin and forms rhodopsin again. Vitamin A is stored in the liver.

Vitamin A plays important roles in reproduction, growth, and epithelial tissue upkeep. Deficiency results in impaired night vision, "night blindness," as well as xerophthalmia. Deficiency may cause the male to be unable to form sperm cells, and may retard growth. Vitamin A maintains epithelial tissues and avoids corneal atrophy.

β-carotene can form 2 retinal molecules, it may also decrease the risk for lung cancer and other carcinomas.

VITAMIN D

Vitamin D Synthesis

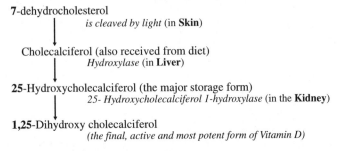

7-dehydrocholesterol
     is cleaved by light (in **Skin**)

Cholecalciferol (also received from diet)
     Hydroxylase (in **Liver**)

25-Hydroxycholecalciferol (the major storage form)
     25- Hydroxycholecalciferol 1-hydroxylase (in the **Kidney**)

1,25-Dihydroxy cholecalciferol
     (the final, active and most potent form of Vitamin D)

Vitamin D maintains calcium levels by:
Increasing resorption of calcium from bone,
Decreasing kidney excretion of calcium,
Increasing the calcium-binding proteins and absorption of calcium (increase calcium uptake) in the intestine.

It increases serum calcium and phosphate levels by bone resorption (in the presence of required parathyroid hormone).

Vitamin D is the most toxic vitamin. A *deficiency* of Vitamin D in children is called Rickett's. A deficiency in adults is called Osteomalacia.

VITAMIN E

Acts as an antioxidant. It is formed of tocopherols and is relatively nontoxic. (*Least* toxic fat-soluble vitamin.) High doses of Vitamin E may decrease the risk for coronary artery disease.

VITAMIN K

Vitamin K is a required vitamin (it acts as a cofactor) for prothrombin (clotting factor 2) and clotting factors 7, 9, and 10 in blood coagulation. Toxicity may cause hemolytic anemia and jaundice. A deficiency may cause a decreased carboxylation of the glutamic acid residues of prothrombin, and a decreased formation of fibrin monomers (from fibrinogen). Also, a deficiency may cause a decrease in the binding of Ca^{++} to prothrombin (factor II). Bacteria in the intestine make Vitamin K except in newborns or antimicrobial therapy, this is a reason newborns receive a shot of Vitamin K.

WATER-SOLUBLE VITAMINS

The remaining 9 vitamins. These vitamins are not considered toxic since the stored levels are low, and at high levels are then excreted in the urine. They include the B vitamins, biotin, pantothenic acid, folic acid, and ascorbic acid.

VITAMIN B_1 (THIAMINE)

Active form: Thiamine pyrophosphate (TPP), acts as a cofactor in oxidative decarboxylation of α-keto acids and by transketolase reactions.

Deficiency can result in Beri-beri and Wernicke-Korsakoff syndrome.

VITAMIN B_2 (RIBOFLAVIN)

Active cofactor forms: Flavin mononucleotide (FMN) and Flavin adenine dinucleotide (FAD).

FAD and FMN bind with hydrogen to form $FADH_2$ and $FMNH_2$. Deficiency results in dermatitis, glossitis, and cheilosis (cracking at the corner of the mouth).

VITAMIN B_6 (PYRIDOXINE)

Pyridoxine, pyridoxamine, and pyridoxal are precursors of active pyridoxal phosphate (a coenzyme). *As a general rule, this vitamin is a co-*

*enzyme in either the reactants or the products when you see **amino acid** reactions—you can feel comfortable to guess "Vitamin B_6" on your exam. Otherwise, memorize the following reactions.* Vitamin B_6 is involved in:

1. Transamination   (OAA + Glu $\longleftrightarrow$ Asp + α-kg)
2. Decarboxylation   (Histidine $\longrightarrow$ CO_2 + Histamine)
3. Deamination   (Serine $\longrightarrow$ NH_3 + Pyruvate)
4. Condensation   (Glycine + SuccinylCoA $\longrightarrow$ δ-ALA)

Deficiency is associated with Isoniazid anti-tuberculosis treatment (therefore, pyridoxine is given). In general, *water*-soluble vitamins are not usually toxic, but vitamin B_6 is actually a toxic vitamin that can result in gait problems from CNS toxicity.

VITAMIN B_{12} (COBALAMIN)

Contains *cobalt* in the center of a corrin ring.
*Methyl*cobalamin contains a methyl group, and *cyano*cobalamin contains CN on a coordinate area.
Involved in the following reactions:

1. Homocysteine — (*methyl*cobalamin) $\longrightarrow$ Methionine

2. Methylmalonyl CoA — (*deoxyadenosyl*cobalamin) $\longrightarrow$ Succinyl CoA

Vitamin B_{12} binds to intrinsic factor for absorption. Deficiency results from decreased parietal cell secretion of intrinsic factor (IF) in the stomach, causing *pernicious megaloblastic anemia*. This is due to decreased IF, and therefore decreased absorption of vitamin B_{12}. Vitamin B_{12} deficiency will also cause *neurologic problems* (folate deficiency only results in megaloblastic anemia).

NIACIN (NICOTINIC ACID)

Active cofactor forms: NAD^+ and $NADP^+$.
They are coenzymes in oxidation-reduction reactions, and are reduced to NADH and NADPH. Tryptophan is able to form niacin. A deficiency results in *Pellagra*: **d**ermatitis, **d**iarrhea, and **d**ementia (the "3-D's").

BIOTIN

An imidazole derivative. Acts as a carrier of carbon dioxide in *carboxylation* reactions. It is found in nearly all food. The only deficiency may appear with eating a diet of raw egg whites (that have avidin). *Avidin* prevents absorption by binding to biotin.

The biotin-requiring reactions that are important to know:
Acetyl CoA $\longrightarrow$ *Biotin* $\longrightarrow$ Malonyl CoA

Pyruvate + HCO_3^- + ATP $\longrightarrow$ *Biotin* + *Acetyl CoA* $\longrightarrow$ OAA + ADP + Pi

FOLIC ACID

Made of p-aminobenzoic acid (PABA) bound to a pterin ring and glutamic acids. The active form is Tetrahydrofolate (THF), which is formed by reducing folate with the enzyme, dihydrofolate reductase (DHF — *dihydrofolate reductase* → THF). PABA analogs (sulfanilamide) will inhibit synthesis of folate. THF transfers carbon for the synthesis of purines, thymidylic acid (a pyrimidine) and amino acids. Deficiency results in *megaloblastic anemia* (frequently seen in alcoholics and pregnant wormen; very common). Pregnant women must receive folic acid *early* in fetal development to avoid *neural tube defects* (i.e., spina bifida) and growth failure. It is recommended to begin vitamin supplements *before* conception.

ASCORBIC ACID (VITAMIN C)

Vitamin C is a reducing agent oxidized by oxygen. It acts as a cofactor in forming hydroxylysyl residues, and is required in the hydroxylation of *collagen*. It helps in the absorption of iron by reducing Ferric (Fe^{3+}) to the Ferrous (Fe^{2+}) form in the stomach. Excretion of increased levels of vitamin C may result in kidney stones by calcium salt deposits of oxalate (a breakdown product). Deficiency results in *scurvy*: deficient hydroxylation of collagen and therefore, connective tissue problems which include sore gums and loose teeth, as well as anemia.

PANTOTHENIC ACID

Part of Coenzyme A, that participates in the transfer of acyl groups. Pantothenic acid is a growth factor that is a part of acyl carrier protein.

Deficiency of	Causes
Vitamin A	Xerophthalmia, Night Blindness
Vitamin D	Rickets in children, Osteomalacia in adults.
Vitamin E	(Usually in *premature* infants)
Vitamin K	(Deficiency is unusual for an adult), Hemorrhagic disease of the newborn, Decreased prothrombin level, Prolonged coagulation time
Vitamin B$_1$ (Thiamine)	Beri-beri (from polished rice as diet) characterized by dry skin and irritability, with eventual death. (therefore, must treat)

Deficiency of	Causes
Vitamin B$_2$ (Riboflavin)	Dermatitis, Glossitis, and Cheilosis
Vitamin B$_6$ (Pyridoxine)	(Associated with Isoniazid anti-tuberculosis treatment)
Vitamin B$_{12}$ (Cobalamin)	Pernicious megaloblastic anemia (from lack of intrinsic factor), Neurologic problems.
Niacin (Nicotinic acid)	Pellagra: "3 D's" *Dermatitis, Diarrhea, and Dementia*
Biotin	Lethargy, acidosis, and dehydration; Increased urine levels of propionic acid, and increased intermediates of isoleucine intermediates. Caused from a diet of raw egg whites (that have avidin)
Folate	Megaloblastic anemia, Neural tube defects in children of deficient pregnant mother.
Ascorbic acid (Vitamin C)	Scurvy (Connective tissue problems; sore gums and loose teeth)
Pantothenic acid	(none to note)

HORMONES

Hormones may be either *polypeptide* hormones or *steroid* hormones. They act as chemical messengers and cause a response in a target tissue or tissues. Hormones have different signaling methods: 1. Hormone will enter the cell and redirect how the gene acts. These are the steroid hormones, Vitamins D and A, and thyroxine. 2. Receptor-mediated will change the shape and signal the interior of the cell. This occurs in the cell *membrane*, NOT inside the cell. It causes a cascade effect. Ion-channels may open up next to the receptor, which allows ions to flow in or out of the cell.

Insulin, an **an**abolic hormone, acts by signaling the *membrane* receptor and activating *tyrosine kinase*. Glucagon acts by changing the shape of the G-proteins and acting on *adenylate cyclase* to convert ATP to cAMP. This excites *protein kinase* which phosphorylates proteins.

INSULIN
Insulin is a peptide hormone made by pancreatic islet β-cells, with **an**abolic effects. It binds to specific receptors in the cell membrane of tissues like liver, muscle, and adipose tissue, and increases the transport of *glucose* and *amino acids* across the plasma membrane. Insulin *increases* protein, triglyceride, and

glycogen *synthesis* in the liver and muscle. By acting on the receptor, it activates or inhibits enzymes in the cell. For example, when insulin signals the membrane receptor, it activates (causes the phosphorylation of) *tyrosine kinase* which then places phosphates on Tyrosine residues of target proteins. Another action of insulin is its ability to decrease the blood concentration of glucose. Furthermore, it decreases the fatty acid level by inhibiting *hormone-sensitive lipase* in fat. As a response to stress, epinephrine will decrease the insulin release. Insulin secretion is *increased* by increased glucose, amino acids, and secretin.

Overall, insulin acts to increase the synthesis of glycogen, protein, and triacylglycerol—it is considered an *anabolic* hormone.

Preproinsulin ⟶ Proinsulin ⟶ Insulin (active form).

GLUCAGON
Glucagon is a peptide hormone that is secreted by pancreatic islet α-cells. It maintains blood glucose by increasing liver glycogenolysis and gluconeogenesis. Glucagon increases blood sugar, breaks down *liver* glycogen, and increases gluconeogenesis (does NOT break down muscle glycogen). It is increased in response to hypoglycemia, to increase the glucose level. Remember, glucagon is *increased* by: a decreased blood glucose level, an increased epinephrine (stress), and a high amino acid diet (a high protein meal needs to counter the hypoglycemic effect that insulin would create). A sudden increase of glucagon would increase liver cAMP levels and liver glycogenolysis. Glucagon increases the transformation of pyruvate to phosphoenolpyruvate. It also increases fructose-1,6-*di*phosphate conversion to fructose-6-phosphate.

OXYTOCIN
Oxytocin causes uterine *contraction*, and the *release* of milk during breast feeding ("milk let down"). This hormone is released by the *posterior* pituitary.

VASOPRESSIN
Antidiuretic hormone (ADH) or vasopressin increases sodium and water reabsorption in the kidney. This hormone is released by the *posterior* pituitary.

GROWTH HORMONE
Growth hormone increases amino acid uptake and growth. Somatotropin is considered a growth hormone. Growth hormone is released by the *anterior* pituitary.

ACTH (ADRENOCORTICOTROPIN HORMONE)
ACTH stimulates growth of the adrenal cortex. This hormone is released by the *anterior* lobe of the pituitary (hypophysis).

PARATHYROID HORMONE
PTH *increases* blood calcium levels. It decreases plasma phosphate level by increasing the *excretion* of phosphate.

CHOLECYSTOKININ (CCK)

CCK is a hormone made by the intestinal endocrine cells of the jejunum and the last part of the duodenum, and is increased in the presence of fat. CCK causes the gallbladder to release bile and the pancreas to release enzymes. CCK decreases motility (less movement of gastric material to the small intestine). This peptide hormone stimulates the contraction of the gallbladder.

SECRETIN

Secretin is released when there is an increase in fats, a low pH (increased gastric hydrochloric acid), and chyme. Secretin causes the pancreas to release *bicarbonate* (pancreatic juice secretion) into the duodenum.

GASTRIN

This hormone is secreted in the pyloric-antral area of the stomach. It increases the HCl secretion of parietal cells. A *gastrinoma* is a gastrin-secreting tumor that is associated with Zollinger-Ellison syndrome.

CALCITONIN

Calcitonin *decreases* blood calcium levels. (*Think of "calcium toning" to decrease the calcium level.*)

TRIIODOTHYRONINE (T_3) AND THYROXINE (T_4)

These thyroid hormones increase heat production and oxygen consumption. *Cretinism* is a prenatal thyroid deficiency that causes defective physical and mental growth.

THYROID STIMULATING HORMONE (TSH)

Activates the secretion of thyroid hormones. Thyroid *stimulating* hormone causes the synthesis and secretion of T_3 and T_4.

STEROID HORMONES

Steroid hormones are lipid soluble hormones that cause effects at a slower rate. This is why you need to give corticosteroids for several days before you see an effect. They decrease the protein synthesis. They activate nuclear receptors of the target tissue and cause an increase in mRNA and protein synthesis to occur.

CALMODULIN

Calmodulin acts as an intracellular receptor for Ca^{2+}. It activates myosin light chain kinase and binds reversibly to Ca^{2+} (in muscle contraction).

ADDITIONAL REVIEW

CATECHOLAMINES

Catecholamines include: Dopamine, Norepinephrine, and Epinephrine. DA and NE are neurotransmitters of the brain. NE and Epinephrine are released from the adrenal medulla. Catecholamines are synthesized from Tyrosine (which is from Phenylalanine):

Tyrosine — 1 ➤ Dopa — 2 ➤ Dopamine — 3 ➤ NE — 4 ➤ Epinephrine

1. Hydroxylase (rate-limiting step; uses tetrahydrobiopterin)
2. Decarboxylase
3. β-Hydroxylase
4. N-Methyltransferase (uses **SAM**)

TETRAHYDROBIOPTERIN (BH_4)

BH_4 is needed in tyrosine, catecholamine, and serotonin synthesis. The following schematic should be memorized:

Phenylalanine — BH_4 ➤ Tyrosine —— BH_4 ➤ Dopa

Tryptophan — BH_4 ➤ 5-Hydroxy tryptamine ➤ 5-HT (Serotonin)

TYPE I DIABETES MELLITUS (INSULIN-DEPENDENT DM, JUVENILE DM)

In type I DM, β-cells are destroyed and there is no insulin production. This occurs as a *child* and can result in *ketoacidosis*. The treatment is insulin to sustain life. Insulin deficiency results in increased glucose level, glucosuria, and ketogenesis.

TYPE II DIABETES MELLITUS (ADULT-ONSET DM)

This occurs as a result of "lazy" β-cells and peripheral resistance of insulin. It occurs in individuals over 35 years old, and is associated with *obesity*. It includes about **90%** of all diabetics. The treatment is *dieting* and then oral hypoglycemics if there is poor compliance or difficulty in controlling the blood sugar levels.

MINERALS

IRON

Iron is **trans**ported in the blood by *transferrin*. *Storage* of extra iron is by *ferritin* and *hemosiderin*. Heme iron is absorbed by mucosal cells of intestine more than non-heme iron. *Primary hemochromatosis* may *accumulate iron* by increased absorption from the diet. Dietary ascorbic acid increases iron absorption. Phosphates bind with iron and decrease absorption. Iron is absorbed mainly in the duodenum and the proximal jejunum.

CALCIUM

The nutritional source for calcium is *milk*. Bones are a reservoir of calcium and they "fill up with calcium" until about 21 years of age. After that time, bones are at a continuous drainage of calcium. Menopausal women are often placed on *estrogen* replacement to decrease the amount of calcium loss and the number of bone fractures. This is why menopausal women have a steep decrease in the calcium and their bones become more fragile. Weight-bearing exercises and activity can help maintain bone calcium.

Biochemical Disorders

8

CHRONIC GRANULOMATOUS DISEASE

Deficiency of *NADPH oxidase*. This results in chronic pyogenic infections since there is no *respiratory burst*.

DIABETES MELLITUS

Insulin deficiency that results in increased glucose level, glucosuria, and can increase ketogenesis. (*See Chap. 7*)

GALACTOKINASE DEFICIENCY

Deficient galactokinase, an autosomal recessive disease, resulting in increased galactose in the urine. Remember that most enzyme deficiencies are *autosomal recessive*.

AMINO ACID DISORDERS

HOMOCYSTINURIA

Increased urine levels of homo**cystine**. May treat with dietary Vitamin B_6 (pyridoxine) and cysteine. Cannot methylate homocysteine into methionine. (Homocysteine is oxidized into homocystine.) Deficiency of cystathionine synthetase.

PHENYLKETONURIA (PKU)

PKU results from a deficiency of phenylalanine hydroxylase. The patient must eliminate phenylalanine from the diet. Tyrosine must be supplied in the diet since it cannot be made from phenylalanine. Neonatal screening test: Guthrie Test may show false negative if done at birth.

ALKAPTONURIA

Deficiency of homogentisate oxidase. Causes darkened urine when standing.

UREA CYCLE DISORDERS

Carbamoyl phosphate synthetase deficiency
Causes hyperammonemia.

Ornithine transcarbamoylase deficiency
Causes hyperammonemia, as well as orotic aciduria.

90

Citrullinemia

Deficiency of argininosuccinate synthetase.
Causes increased citrulline level.

Argininemia

Deficiency of argininosuccinase.
Causes argininosuccinate excretion in the urine.

MAPLE SYRUP URINE DISEASE

This disease was named because of the "sweet urine."
Deficiency of keto-acid (branched-chain) dehydrogenase.
Therefore, branched-chain keto acids in urine.
(branched-chain amino acids are isoleucine, leucine, and valine)
To treat, the diet must be low in branched-chain amino acids.

GLYCOGEN STORAGE DISORDERS (12 TYPES OF GSD)

(Most are *autosomal recessive*, with the exception of phosphorylase b
kinase deficiency)

VON GIERKE'S (TYPE I GSD)

Deficiency of glucose-6-phosphatase. Hepatomegaly is seen.

POMPE'S (TYPE II GSD)

Deficiency of α-1,4-glucosidase and lysosomal acid maltase (a debranching enzyme).
Cardiomegaly, restrictive cardiomyopathy, occurs as the heart stores
glycogen.

CORI'S

Deficiency of the debranching enzyme, amylo-1,6-glucosidase.
Effects limit dextrin conversion to glucose.

McARDLE'S (TYPE V GSD)

Deficiency of *muscle* phosphorylase.
Increased glycogen since can't break down glycogen to glucose.
There is weakness and cramps in the muscle.

HERS

Deficiency of *liver* phosphorylase.
Increased glycogen.

PHOSPHORYLASE B KINASE DEFICIENCY

Deficiency of liver phosphorylase b kinase.
Increased glycogen due to inability to activate to the phosphorylase a.

LIPID DISORDERS

FAMILIAL HYPERCHOLESTEROLEMIA

Mutated LDL receptors that reduce binding of LDL.
Increased level of cholesterol.

FAMILIAL LIPOPROTEIN LIPASE DEFICIENCY
Deficiency of lipoprotein lipase
Increased levels of chylomicron triglyceride.

FAMILIAL LCAT DEFICIENCY
Deficiency of lecithin: cholesterol acyltransferase.

TANGIER'S DISEASE
Defective HDL synthesis.
Elevated cholesterol in tissue and a decreased plasma HDL level.

FAMILIAL DYSBETALIPOPROTEINEMIA
Deficiency of Apo E 3.

ABETALIPOPROTEINEMIA
Defective synthesis of Apo B-lipid complex.

LYSOSOMAL STORAGE DISEASES

Lysosomes have hydrolytic enzymes. When these enzymes decrease, there is an accumulation of sphingolipids and sulfates, etc.

MUCOPOLYSACCHARIDOSES

Hunter's
X-linked recessive deficiency of iduronosulfate sulfatase.
This causes an increase in dermatan and heparan sulfate.
Causes mental and physical retardation. (NO corneal clouding)

Scheie's
Autosomal recessive deficiency of α-L-iduronidase.
Causes clouding of the cornea, joint degeneration, and heart disease. These individuals have a *normal* life expectancy.
Does NOT cause retardation.

Hurler's
Deficiency of α-L-iduronidase.
Increased dermatan and heparan sulfate.
Causes clouding of the cornea, mental and physical retardation, and early death.

Sanfilippo's
Deficiency by Type:
A: N-sulfatase,
B: N-acetylglucosaminidase,
C: N-acetyl CoA: α-glucosamine-acetyltransferase,
D: N-acetyl-α-D-glucosamine-6-sulfatase.
This increases heparan sulfate
Causes severe mental retardation.
Type III mucopolysaccharidosis.

SPHINGOLIPIDOSES

GM1-gangliosidosis
Defective β-gangliosidase A.

Tay-Sachs Disease
Deficiency of hexosaminidase A.
Increased GM2-gangliosides.
This disease is associated with Ashkenazi (Eastern European) Jews, and a *macular cherry-red spot.*
Causes retardation and early death.

Gaucher's
Deficiency of glucocerebrosidase.
Increased glucocerebrosides.

Niemann-Pick's
Deficiency of sphingomyelinase.
Increased sphingomyelin.
This is associated with severe mental retardation and early death.

Fabry's
Deficiency of α-galactocerebrosidase A.
Increased ceramide trihexosides.
This is the only sex-linked recessive *sulfatidosis.*

PURINE AND PYRIMIDINE DISORDERS

LESCH-NYHAN
Deficiency of HGPRT (but, has xanthine oxidase).
Self-mutilating disease that occurs only in males.
Lesch-Nyhan Syndrome is *associated with:*
Gout, increased serum urate,
increased **p**hospho**r**ibosyl**p**yro**p**hosphate (PRPP), increased hypoxanthine (due to HGPRT deficiency), self-destruction, and mental retardation.

Allopurinol (the treatment for *gout*) is an analogue of xanthine and is used to *inhibit xanthine oxidase*, this decreases hypoxanthine conversion to urate. (But, still have neuro problems.) Allopurinol inhibits uric acid production and urate crystal formation.

Xanthine oxidase is involved in the degradation of purines; two oxidation reactions:
Hypoxanthine $\longrightarrow$ Xanthine $\longrightarrow$ Uric acid.
(Xanthine is also a precursor of guanine)

ADENOSINE DEAMINASE (ADA) DEFICIENCY
Defect in gene for adenosine deaminase, leads to severe combined immunodeficiency disease (SCID).

Remember, Adenosine deaminase was the first gene transplanted into human cell DNA to develop the immune system. Adenosine is converted into Inosine (*See purine nucleotide degradation*). An adenosine deficiency will shut off deoxyribose production, and this decreases the immune system. Think of the "Bubble baby."

OROTIC ACIDURIA
Deficiency of orotate phosphoribosyltransferase.
Causes megaloblastic anemia and increased orotate excretion in the urine.

XERODERMA PIGMENTOSUM
Deficiency of U.V. endonuclease.
Result of the loss of excision-repair to remove thymine dimers (created by UV radiation forming dimers in DNA).
Repair dimers by 3 enzymes:
1. Specific exonuclease-protein complex
2. DNA polymerase I
3. DNA ligase

Ligase is a repair enzyme of thymine dimers from excessive sun exposure. It is involved in DNA repair and synthesis, and in the formation of phosphodiester bonds between dsDNA molecules. (*See Chap. 4*)

OTHER DISORDERS

WILSON'S DISEASE
Increased serum copper due to deficiency of ceruloplasmin.
Causes hepatolenticular degeneration as well as dermatitis.

ACUTE INTERMITTENT PORPHYRIA
Deficient porphobilinogen deaminase.
Increased levels of δ-aminolevulonic acid (δ-ALA)

CRIGLER-NAJJAR SYNDROME
Defective formation of bilirubin glucuronide.
Deficiency of hepatic bilirubin-UDP-glucuronyl transferase.
Nonhemolytic jaundice, hyperbilirubinemia.
May cause irreversible brain damage.

HEMOSIDEROSIS
Excessive iron storage,
(Hemochromatosis),
Can occur if receive many blood transfusions.

HEMOGLOBINOPATHIES

Sickle cell anemia
Substitution of **valine** for glutamate at position **6** on β-chains in hemoglobin.

Abnormal Hemoglobin S.
The **de**oxygenated hemoglobin is *less* soluble than the oxygenated hemoglobin.
Electrophoresis demonstrates Hemoglobin S band.
(Sickle cell *trait* demonstrates both hemoglobins S and A)

Thalassemias
Cause defective synthesis (abnormal quantity) of α- and β-hemoglobin chains.
Normal function of the chains.
α-thalassemia patient may produce hemoglobin H; decreased α-chain synthesis.
β-thalassemia has decreased β-chain synthesis.
β-thalassemia may be caused by a gene deletion or a mutation that may affect the processing of mRNA or premature chain termination.

CONNECTIVE TISSUE DISORDERS

Ehlers-Danlos Syndrome
Collagen defects.
Can occur with decreased lysyl oxidase or lysyl hydroxylase.
Elastic skin and loose joints. These individuals are "bendable."

Marfan's Syndrome
Deficiency of type I collagen formation. Elastin defect.
Tall stature, arachnodactyly (long spider fingers).
Weak arteries can lead to a dissecting aortic aneurysm.

Osteogenesis Imperfecta
Decreased collagen synthesis.
"Brittle bone syndrome," bones bend and break.

HARTNUP DISEASE
Transport defect of tryptophan in intestinal and renal systems.

CYSTINURIA
Defective cystine, lysine, arginine, and ornithine transport.
Cystine crystals and stone formation.

BLOOD COAGULATION DEFICIENCIES

Classic hemophilia
Factor 8 deficiency or Hemophilia A. X-linked recessive.
These individuals have problems with blood coagulation and will have massive hemorrhage after trauma or an operation. They have *normal* bleeding time, but a *prolonged* coagulation time.
Treatment: Factor VIII concentrate.

Christmas disease
Factor 9 deficiency or Hemophilia B. X-linked recessive.

Disorder	Accumulation in urine of:
Phenylketonuria (PKU)	Phenylpyruvate, Phenylalanine
Maple syrup urine disease	Keto acids
Homocystinuria	Homocysteine
Histidinemia	Histidine
Alkaptonuria	Homogentisate polymers; (darkened urine on standing)

Disorder	Deficiency	Clinical Presentation
Chronic granulomatous dis.	NADPH oxidase	
Diabetes Mellitus	Insulin	*Increased:* Glucose level, glucosuria, and possible ketogenesis
Galactokinase deficiency	Galactokinase	Increased galactose in urine
Homocystinuria	Cystathionine synthetase	Increased homocytine in urine
Phenylketonuria	Phenylalanine hydroxylase	Must be given tyrosine in diet, Must eliminate phenylalanine from diet.
Alkaptonuria	Homogentisate oxidase	Darkened urine when stand
Carbamoyl phosphate synthetase deficiency	Carbamoyl phosphate synthetase	Hyperammonemia
Ornithine transcarbamoylase deficiency	Ornithine transcarbamoylase	Hyperammonemia, orotic aciduria
Argininemia	Argininosuccinase	Increase urine excretion of Argininosuccinate

Disorder	Deficiency	Clinical Presentation
Maple syrup urine disease	Keto acid dehydrogenase	Increased urine ketoacids. (branched chain A.A.'s) Sweet smell of urine.
Von Gierke's	Glucose-6-phosphatase	Hepatic glycogenosis.
Cori's	Debranching enzyme, amylo-1,6-glucosidase	
McArdle's	*Muscle* phosphorylase	Increased glycogen
Hers	*Liver* phosphorylase	"
Phosphorylase b kinase defic.	Liver phosphorylase b kinase	"
Familial hypercholesterolemia	LDL receptors mutated	Increased cholesterol
Familial lipoprotein lipase deficiency	Lipoprotein lipase	Increased chylomicron TG
Familial LCAT deficiency	Lecithin: cholesterol acyltransferase	
Tangier's disease	Defective HDL synthesis	Increased cholesterol, Decreased HDL level
Familial dysbeta-lipoproteinemia	Apo E 3	
Abetalipoproteinemia	Defective synthesis of Apo-B lipid complex	
Hunter's	Iduronate sulfatase	Mental and physical retardation. NO corneal clouding
Scheie's	α-L-iduronidase	Joint degeneration, corneal clouding. NO retardation
Hurler's	α-L-iduronidase	Mental and physical retardation, corneal clouding. Early death.

Disorder	Deficiency	Clinical Presentation
Sanfilippo's	Type A: N-sulfatase Type B: N-acetylgluco-saminidase	Severe mental retardation "Type III mucopoly-saccharidosis"
	Type C: N-acetyl CoA: α-glucosamine-acetyltransferase	Hepatomegaly
	Type D: N-acetyl-α-D-glucosamine-6-sulfatase	
GM1-gangliosidosis	β-gangliosidase A	
Tay-Sachs disease	Hexosaminidase A	Retardation, early death. Cherry-red spots from deposits of N-acetylgalacto-samine in the retina and brain. GM2 gangliosides.
Gaucher's	Glucocerebrosidase	
Niemann-Pick's	Sphingomyelinase	Also, Cherry-red spots.
Fabry's	α-galactosidase A	Skin and kidney accumulation of ceramide trisaccharides.
Lesch-Nyhan Syndrome	HGPRT (totally absent)	Gout, Mental retarda-tion, self-destruc-tion. Increased: serum urate, phos-phoribosylpyro-phosphate (PRPP), hypoxanthine.
Gout	HGPRT is decreased	Urate crystals.
ADA deficiency	Adenosine deaminase	SCID
Orotic aciduria	Orotate phospho-ribosyltransferase	Megaloblastic anemia Increased urine excretion of orotate

Disorder	Deficiency	Clinical Presentation
Xeroderma Pigmentosum	UV endonuclease	Skin damage from sun exposure
Wilson's Disease	Ceruloplasmin	Increased serum copper, Hepatolenticular degeneration, Dermatitis.
Acute Intermittent Porphyria	Porphobilinogen deaminase	Increased δ-ALA
Crigler-Najjar Syndrome	Hepatic bilirubin-UDP-glucoronyl transferase	Jaundice, Hyperbilirubinemia, (unconjugated bilirubin) Irreversible brain damage may occur.
Hemosiderosis		Excessive iron storage
Sickle cell anemia	(Caused by a point mutation that substitutes *valine* for glutamate at position 6 on β-chains in Hb)	Hemolysis, Ulcers, Infarcts of bone and spleen. Autosplenectomy.
Thalassemia	α- and β-hemoglobin chains	Hemoglobinopathy
Ehlers-Danlos Syndrome	Lysyl oxidase, Lysyl hydroxylase	Collagen defects
Classic Hemophilia	Factor 8	Blood coagulation problems, Bleeding.
Christmas Disease	Factor 9	Blood coagulation problems, Bleeding.
I-Cell Disease	Mannose phosphorylation is decreased	Skeletal abnormality, Childhood death.

Miscellaneous 9

SUMMARY OF ENERGY PRODUCTION AND CONSUMPTION

	Production	Consumption	Net Gain	Net Loss
Glycolysis (anaerobic; Substrate level)	4 ATP	2 ATP	+ 2 ATP	
Glycolysis (aerobic; Oxidative phosphorylation)	2 ATP 2 NADH	2 ATP	+ 6 ATP	
Gluconeogenesis (from pyruvate)		2 ATP 2 GTP 2 NADH		2 ATP 2 ATP 6 ATP −10 ATP
Kreb's Cycle (substrate level) (oxidative phosphorylation) (FAD oxidation)	2 GTP 6 NADH 2 FADH$_2$		2 ATP 18 ATP 4 ATP +24 ATP	
Pentose Phosphate Path (oxidative)	2 NADPH			

1 GTP + 1 ADP = 1 GDP + 1 ATP

Then, through the electron chain:
1 NADH = 3 ATP
1 FADH$_2$ = 2 ATP

Hormonal Control of Metabolism

Hormone	Increases	Decreases
Insulin	Glycogen synthesis Protein synthesis Lipid synthesis Transport of glucose into some tissues, i.e., muscle.	Gluconeogenesis Glycogenolysis (blood sugar level) Lipolysis Protein degradation
Glucagon	Gluconeogenesis Glycogenolysis Lipid breakdown (blood sugar level)	Glycolysis

S-ADENOSYLMETHIONINE (SAM)

Donates methyl group.

SAM ←——→ S-adenosyl*homocysteine*,

S-adenosyl*homocysteine* ←——→ Homocysteine + Adenosine.

Reactions involving SAM: (Donates methyl group)

Phosphatidyl*ethanolamine* ——→ Phosphatidyl*choline*,

Guanidinoacetic acid ——→ Creatine,

Guanine ——→ Terminal 7-methylguanylate cap (5'-cap),

Norepinephrine ——→ Epinephrine.

Post-transcriptional modification reactions
(i.e., 1-*methyl*adenine in tRNA)

S-Adenosylmethionine

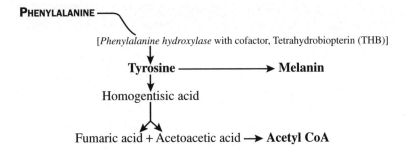

PHENYLALANINE

[*Phenylalanine hydroxylase* with cofactor, Tetrahydrobiopterin (THB)]

Tyrosine ⟶ **Melanin**

Homogentisic acid

Fumaric acid + Acetoacetic acid ⟶ **Acetyl CoA**

FUMARATE AND ASPARTATE
Connect the citric acid cycle with the urea cycle

UREA CYCLE

Nitrogenous waste products are removed through the urea cycle. Urea is formed in the liver, enters the blood and is excreted in the urine. One nitrogen is from ammonium (NH_4^+ via carbamoyl phosphate), the other is from aspartate. Reactions occur in the cytosol and mitochondrial matrix.

First two reactions occur in the mitochondria:

1. Carbamoyl phosphate synthetase reaction
   $$2\,ATP + CO_2 + NH_4^+ + H_2O \longrightarrow Carbamoyl\ phosphate + 2\,ADP + P_i$$

2. Ornithine carbamoyl transferase reaction
   $$Ornithine + Carbamoyl\ phosphate \longrightarrow Citrulline + P_i$$

The remaining reactions occur in the cytosol:

3. Argininosuccinate synthetase reaction
   $$Citrulline + \textbf{Aspartate} + ATP \longrightarrow Argininosuccinate + AMP + PP_i$$

4. Argininosuccinate lyase reaction
   $$Argininosuccinate \longrightarrow Fumarate + Arginine$$

5. Arginase reaction
   $$Arginine \longrightarrow Ornithine + \textbf{UREA}$$

(Ornithine returns into the mitochondria. While, the toxic NH_3 is released as urea.)

Urea:

$$NH_2{-}\underset{\underset{O}{\|}}{C}{-}NH_2$$

4 moles of high energy phosphate bonds are released in the synthesis of 1 mole of urea.

Urea Cycle

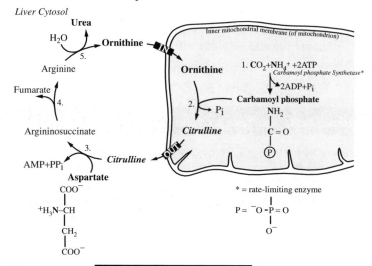

AMINO ACID CATABOLISM
(The use of carbons in metabolic pathways):

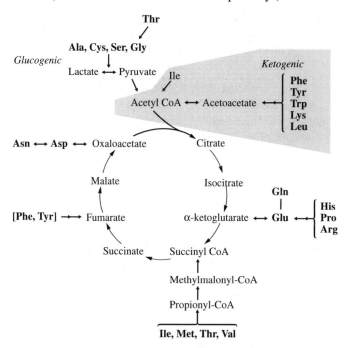

Threonine → Glycine → Serine → Pyruvate

Threonine → Propionyl CoA → Methylmalonyl CoA → Succinyl CoA

Homocysteine — (*methyl*cobalamin) → Methionine

Aspartic acid can enter the TCA cycle or the urea cycle (NH_4^+).
Transamination of aspartate will convert to oxaloacetate.

aspartate aminotransferase
Aspartate + α-ketoglutarate ⟷ Oxaloacetate + Glutamate

Amino Acids and Kreb's Cycle

Purely Ketogenic	Glucogenic	Both Keto/Glucogenic
Leucine Lysine	(All others form glucose)	Isoleucine Phenylalanine Tryptophan Tyrosine

NUCLEOTIDE METABOLISM

Nucleotides are important for the synthesis of lipids, carbohydrates, and proteins. They serve in the formation of cofactors and are crucial for metabolic pathways. The two types of bases: purines (A, G) and pyrimidines (C, U, T), are synthesized *de novo* or by salvage pathways.

PURINE NUCLEOTIDE SYNTHESIS

DE NOVO SYNTHESIS
(Adenine and Guanine)
Purine ring structure has several sources for its atoms:

Purine Structure
(Adenine and Guanine)

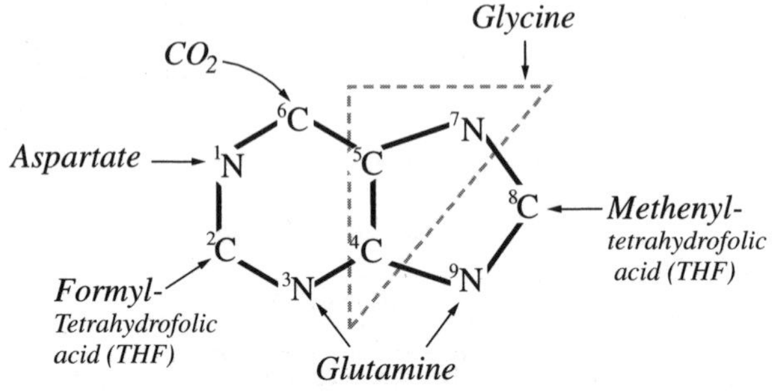

PURINE NUCLEOTIDE SYNTHESIS *(continued)*

5-phosphoribosyl-1-pyrophosphate (PRPP)

$$\text{Ribose-5-phosphate} + \text{ATP} \xrightarrow[\text{ribosephosphate pyrophosphokinase}]{} \text{PRPP}$$

Then, the committed step:

$$\text{PRPP} + \text{Glutamine} \xrightarrow[]{\text{Glutamine phosphoribosyl pyrophosphate aminotransferase}} \text{5-phospho-}\beta\text{-D-ribosylamine} + \text{Glutamate}$$

Several steps proceed to form Inosine monophosphate (IMP)

IMP is an *unusual* base found in tRNA. Glycine, Methenyl-THF, Glutamine, ATP, CO2, Aspartate, and Formyl-tetrahydrofolate enter the path to form IMP.

Finally, IMP is converted to AMP and GMP

$$\text{IMP} + \text{Aspartic acid} + \textbf{GTP} \xrightarrow[]{\textit{adenylosuccinate synthase}} \text{Adenylosuccinate} + \text{GDP} + \text{P}_i$$

$$\text{Adenylosuccinate} \xrightarrow[]{} \textbf{AMP} + \text{Fumarate}$$

$$\text{IMP} + \text{NAD}^+ \xrightarrow[]{\textit{IMP dehydrogenase}} \text{Xanthosine monophosphate} + \text{NADH} + \text{H}^+$$

$$\text{Xanthosine monophosphate} + \text{Glutamine} + \textbf{ATP} \xrightarrow[]{}$$
$$\textbf{GMP} + \text{Glutamate} + \text{AMP} + \text{PP}_i$$

(As AMP increases it will inhibit adenylosuccinate synthase and as GMP increases it will inhibit IMP dehydrogenase; resulting in decreased conversion of IMP to AMP and GMP)

PURINE NUCLEOTIDE DEGRADATION

The final degradation product of the purine nucleotides is *uric acid*, which is excreted in the urine.

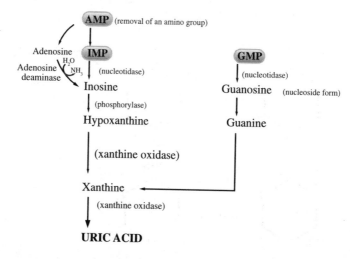

ALLOPURINOL

This drug inhibits *xanthine oxidase*, and is useful in the treatment of Lesch-Nyhan gout.

THE SALVAGE PATHWAY OF PURINES

This path allows undegraded purines from the cellular turnover or diet to be reused by the body. Unlike de novo, this path occurs in *extra*hepatic tissues. Adenine phosphoribosyltransferase and hypoxanthine-guanine phosphoribosyltransferase (HGPRT) are the two enzymes involved. PRPP is used by these enzymes for the ribose-5-phosphate group. Clinical correlation: Lesch-Nyhan Syndrome is the result of HGPRT deficiency, is X-linked recessive, and leads to increased uric acid and self-mutilation.

PYRIMIDINE SYNTHESIS

Synthesis of the pyrimidine ring occurs before attachment to ribose-5-phosphate. PRPP is the donor for the ribose-5-phosphate. Aspartate transcarbamoylase is the enzyme for the committed step.

Pyrimidine ring sources:
Formation of the pyrimidine ring from carbamoyl phosphate and aspartic acid.

Pyrimidine Synthesis

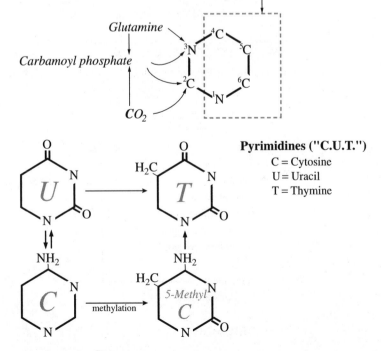

Pyrimidines ("C.U.T.")
C = Cytosine
U = Uracil
T = Thymine

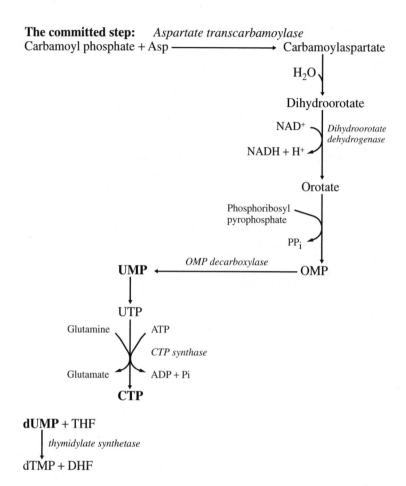

The committed step: *Aspartate transcarbamoylase*
Carbamoyl phosphate + Asp ⟶ Carbamoylaspartate

THF gives up a carbon unit and a hydrogen atom, resulting in DHF. DHF is reduced to THF with dihydrofolate reductase. This enzyme is inhibited by aminopterin and methotrexate and therefore decrease the amount of THF. The reduced THF causes a decrease in the purine synthesis and the inability to methylate dUMP to dTMP. The result: inhibited DNA synthesis.

The end product—UMP, (as well as AMP) inhibit pyrimidine synthesis. ATP, and PRPP activate synthesis. Kinases form diphosphates from monophosphates.

ASPARTATE TRANSCARBAMOYLASE
Controls rate of *Pyrimidine* synthesis (C, U, T).
Regulated by allosteric interaction.
Inhibited by CTP. (End product of pyrimidine synthesis.)

PYRIMIDINE NUCLEOTIDE DEGRADATION

Pyrimidine rings are able to open and can be further used by the body.

Thymine —▶ Dihydrothymine —▶ Ureidoisobutyrate

Uracil ◀—————————▶ *β-aminosobutyrate*

FORMING RIBONUCLEOTIDES FROM DEOXYRIBONUCLEOTIDES

Ribonucleoside reductase is specific for the diphosphates:
ADP, CDP, GDP, UDP.

Thioredoxin donates the hydrogen atoms to reductase enzyme and reduces the ribonucleotide to *deoxy*ribonucleotide. Later, NADPH + H$^+$ and thioredoxin reductase can convert the thioredoxin back to its reduced form and repeat its function.

Congratulations, you have reviewed all of the important topics in Biochemistry!

Index

ISBN 0-07-038217-4

90000